Topics in Engineering Mathematics

Mathematics and Its Applications

Managing Editor:

M. HAZEWINKEL

Centre for Mathematics and Computer Science, Amsterdam, The Netherlands

Volume 81

Topics in Engineering Mathematics

Modeling and Methods

edited by

Adriaan van der Burgh

and

Juriaan Simonis

Faculty of Technical Mathematics and Informatics,
University of Technology,
Delft, The Netherlands

SPRINGER-SCIENCE+BUSINESS MEDIA, B.V.

Library of Congress Cataloging-in-Publication Data

Topics in engineering mathematics : modeling and methods / edited by
Adriaan van der Burgh and Juriaan Simonis.
      p.   cm. -- (Mathematics and its applications ; v. 81)
   ISBN 978-94-010-4800-2   ISBN 978-94-011-1814-9 (eBook)
   DOI 10.1007/978-94-011-1814-9
   1. Engineering mathematics. 2. Mathematical models.   I. Burgh,
Adriaan Herman Pieter van der. II. Simonis, Juriaan, 1943-   .
III. Series: Mathematics and its applications (Kluwer Academic
Publishers) ; v. 81.
TA330.T67 1992
620'.001'51--dc20   92-33608

ISBN 978-94-010-4800-2

Printed on acid-free paper

SERIES EDITOR'S PREFACE

'Et moi, ..., si j'avait su comment en revenir, je
n'y serais point allé.'

Jules Verne

The series is divergent; therefore we may be
able to do something with it.

O. Heaviside

One service mathematics has rendered the
human race. It has put common sense back
where it belongs, on the topmost shelf next to
the dusty canister labelled 'discarded nonsense'.

Eric T. Bell

Mathematics is a tool for thought. A highly necessary tool in a world where both feedback and nonlinearities abound. Similarly, all kinds of parts of mathematics serve as tools for other parts and for other sciences.

Applying a simple rewriting rule to the quote on the right above one finds such statements as: 'One service topology has rendered mathematical physics ...'; 'One service logic has rendered computer science ...'; 'One service category theory has rendered mathematics ...'. All arguably true. And all statements obtainable this way form part of the raison d'être of this series.

This series, *Mathematics and Its Applications*, started in 1977. Now that over one hundred volumes have appeared it seems opportune to reexamine its scope. At the time I wrote

"Growing specialization and diversification have brought a host of monographs and textbooks on increasingly specialized topics. However, the 'tree' of knowledge of mathematics and related fields does not grow only by putting forth new branches. It also happens, quite often in fact, that branches which were thought to be completely disparate are suddenly seen to be related. Further, the kind and level of sophistication of mathematics applied in various sciences has changed drastically in recent years: measure theory is used (non-trivially) in regional and theoretical economics; algebraic geometry interacts with physics; the Minkowsky lemma, coding theory and the structure of water meet one another in packing and covering theory; quantum fields, crystal defects and mathematical programming profit from homotopy theory; Lie algebras are relevant to filtering; and prediction and electrical engineering can use Stein spaces. And in addition to this there are such new emerging subdisciplines as 'experimental mathematics', 'CFD', 'completely integrable systems', 'chaos, synergetics and large-scale order', which are almost impossible to fit into the existing classification schemes. They draw upon widely different sections of mathematics."

By and large, all this still applies today. It is still true that at first sight mathematics seems rather fragmented and that to find, see, and exploit the deeper underlying interrelations more effort is needed and so are books that can help mathematicians and scientists do so. Accordingly MIA will continue to try to make such books available.

If anything, the description I gave in 1977 is now an understatement. To the examples of interaction areas one should add string theory where Riemann surfaces, algebraic geometry, modular functions, knots, quantum field theory, Kac-Moody algebras, monstrous moonshine (and more) all come together. And to the examples of things which can be usefully applied let me add the topic 'finite geometry'; a combination of words which sounds like it might not even exist, let alone be applicable. And yet it is being applied: to statistics via designs, to radar/sonar detection arrays (via finite projective planes), and to bus connections of VLSI chips (via difference sets). There seems to be no part of (so-called pure) mathematics that is not in immediate danger of being applied. And, accordingly, the applied mathematician needs to be aware of much more. Besides analysis and numerics, the traditional workhorses, he may need all kinds of combinatorics, algebra, probability, and so on.

In addition, the applied scientist needs to cope increasingly with the nonlinear world and the extra mathematical sophistication that this requires. For that is where the rewards are. Linear models are honest and a bit sad and depressing: proportional efforts and results. It is in the nonlinear world that infinitesimal inputs may result in macroscopic outputs (or vice versa). To appreciate what I am hinting at: if electronics were linear we would have no fun with transistors and computers; we would have no TV; in fact you would not be reading these lines.

There is also no safety in ignoring such outlandish things as nonstandard analysis, superspace and anticommuting integration, p-adic and ultrametric space. All three have applications in both electrical engineering and physics. Once, complex numbers were equally outlandish, but they frequently proved the shortest path between 'real' results. Similarly, the first two topics named have already provided a number of 'wormhole' paths. There is no telling where all this is leading - fortunately.

Thus the original scope of the series, which for various (sound) reasons now comprises five subseries: white (Japan), yellow (China), red (USSR), blue (Eastern Europe), and green (everything else), still applies. It has been enlarged a bit to include books treating of the tools from one subdiscipline which are used in others. Thus the series still aims at books dealing with:

- a central concept which plays an important role in several different mathematical and/or scientific specialization areas;
- new applications of the results and ideas from one area of scientific endeavour into another;
- influences which the results, problems and concepts of one field of enquiry have, and have had, on the development of another.

The shortest path between two truths in the real domain passes through the complex domain.

J. Hadamard

La physique ne nous donne pas seulement l'occasion de résoudre des problèmes ... elle nous fait pressentir la solution.

H. Poincaré

Never lend books, for no one ever returns them; the only books I have in my library are books that other folk have lent me.

Anatole France

The function of an expert is not to be more right than other people, but to be wrong for more sophisticated reasons.

David Butler

Bussum, 14 August 1992

Michiel Hazewinkel

To
the academic staff of
Jurusan Matematika
INSTITUT TEKNOLOGI SEPULUH NOPEMBER
SURABAYA, INDONESIA

Contents

Preface

Within the framework of a cooperation project in basic sciences between the Department of Mathematics of the Institute of Technology in Surabaya (Indonesia) and the Faculty of Technical Mathematics and Informatics of the University of Technology in Delft (The Netherlands) a symposium on Topics in Engineering Mathematics was planned to be held from 29 June - 4 July 1992 in Surabaya, Indonesia.

On request of the Indonesian Government, The Netherlands terminated its role as a donor country on 25 April 1992. As a consequence the symposium was cancelled. However, when it became clear that the symposium could not take place, the first drafts of the Dutch contributions were already available. The organizers of this symposium strongly feel that this unique collection of expository papers should not be lost and hence decided to publish these papers in a special volume. It is hoped that in this way the contributions, ranging from Karmarkar's method to the four-wheel steering problem in car industry, will find their way not only to the universities in Indonesia but also to all the non-specialists who are interested in recent developments in engineering mathematics.

Finally, it is a pleasure to thank the contributors who responded so positively to the request to prepare these special papers, all referees for their valuable advice and their understanding for the rather tight deadlines and, last but not least, Suharmadi Sanjaya, Head of the Mathematics Department of the Institute of Technology in Surabaya and J.P. Dotman, the Dutch resident engineer, who did a lot of local preparatory work for the symposium.

Delft, July 1992

A.H.P. van der Burgh
J. Simonis

A kaleidoscopic excursion into numerical calculations of differential equations

E. van Groesen
Faculty of Applied Mathematics
University of Twente
P.O. Box 217, 7500 AE Enschede, The Netherlands

Abstract

In this expository paper a sketch is given of some basic problems that arise when differential equations are discretised and calculated numerically. Some analytic techniques are demonstrated for the investigation of these numerical schemes. Numerous pictures illustrate the ideas and the formulae. References to available software packages are given.

For the logistic equation the time asymptotic behaviour depends on the chosen time step, showing period doubling from a correct attraction to the stable equilibrium solution to chaotic behaviour. For the linear harmonic oscillator energy conservation for large times is investigated.

Besides these ordinary differential equations, several aspects of linear and nonlinear wave equations, described by partial differential equations, will also be considered. A finite difference method is demonstrated for a linear wave equation. For the Korteweg-de Vries equation, an equation that combines dispersive and nonlinear effects in wave propagation, Fourier truncations are studied. It is shown that the basic Hamiltonian structure is preserved (implying energy conservation), as well as a continuous translation symmetry (implying conservation of horizontal momentum). As a consequence, travelling waves - i.e. wave profiles that translate with constant speed undeformed in shape- are present in truncations of any order. Numerical evidence, as well as a complete analytical proof, is given.

A preliminary version of a software package WAVEPACK, aimed to familiarize the unexperienced user with many basic concepts from wave theory (dispersion, groupvelocity, mode analysis, etc.) and to perform actual calculations in an easy way, is available upon request.

Keywords: model consistent discretisations, logistic equation, harmonic oscillator, wave equations, Korteweg - de Vries equation, travelling waves.
1991 Mathematics Subject Classification: 65-02, 35Q20.

1

A. van der Burgh and J. Simonis (eds.), Topics in Engineering Mathematics, 1–36.
© 1992 *Kluwer Academic Publishers.*

1   Introduction

This expository paper has several goals. It is written by an applied analysist who occasionally feels the need to use the computer, for instance to produce pictures or to get an idea of the behaviour of some set of differential equations. Mostly I use simple standard software packages, many of which are available nowadays and operate on PC's; incidentially I need the help of numerical analysists to produce my own simple programmes. All the results and pictures in this paper were obtained in either way, and can be reproduced by anyone who is interested.

When performing such calculations, this goes along with a considerable amount of scepsis from my part: do I get (from the computer) what I really wanted, or is the calculated phenomenon a consequence of some unforeseen property of the numerical scheme that is used for the computations. (Even if "you get what you see", it may not be true that what you see is what you should see.) My experience from simple examples is that the latter thing often happens. This explains my scepsis, and at the same time increases my respect for scientists who are able to calculate turbulent flows, the flow around wings, the atmospheric conditions for wheather forecasts, to mention just a few of the most difficult areas of application for numerical calculations. Of course, scientists involved are very much aware of the many pittfalls, and try to cope with them as good as possible. Because, despite the difficulties, numerical calculations are extremely usefull, as shown by the results of these aplications.

My aim is to demonstrate with much simpler examples some of the fundamental problems that are encountered. And how they sometimes (partly) can be resolved by exploiting some specific knowledge one has about the problem to be calculated. All examples deal with differential equations, both ordinary and partial differential equations. We will look at these problems, and their discretisation, with the eye of an analysist, and use some basic analytical techniques for their investigation. The text should be understandable for advanced students (math, physics, engineering) who are willing to put enough effort into the subject to grasp the relevance of the reasoning and the concepts used. Even if a few of the more difficult concepts are not understood completely, this should not hinder to follow the main line of the rest of the text.

In the remainder of this introduction we will make some of the general ideas more explicit.

Continuous variables

As stated, we will deal with problems connected to numerical computations. Such computations were in fact the first and primary aim computers were designed for: to make large calculations to solve problems that are encountered in science and in technical applications. A fundamental problem arises because of the discrepancy between the natural, simplest, way to describe these phenomena in mathematical models, and the kind of problems a computer can handle: it is the difference between a continuous and a discrete description. Let us look at this in somewhat more detail.

Physics, Chemistry, and the natural sciences in general, provide the basic laws that describe the phenomena of nature. They do so, more or less accurate depending on the particular phenomenon or situation under investigation.

In many cases, these laws of nature are formulated in variables describing the "state" of the system. These variables are considered, or idealised, to be continuous. For instance, to describe a fluid in a certain spatial domain, it is in many instances convenient to pretend as if the fluid does not consist of individual molecules, but as if the important quantities like mass and velocity are smoothed out over the whole domain. Therefore the continuum description uses concepts like mass density (mass per volume) and fluid velocity (summed velocity of the particles per volume) that are defined at *each* point by a limiting procedure over ever decreasing volumes around the point under consideration. Usually, a continuum description leads to equations that can be formulated very compactly in mathematical terms, often by a set of (partial) differential equations.

In dynamical problems, whether continuous or discrete in space, the evolution of the system is described with the time, which is a continuous variable by its nature. Hence all these problems are of a continuous nature.

Calculations for such "continuous" problems can be performed by a computer provided they are reformulated in algorithms which break the computations into a few elementary operations that the hardware can handle. Basically, a computer can work only with a finite number of quantities. Although that number is quite large in modern computers, the number is essentially finite. Therefore, the computer requires that the variables and the equations are "discretised".

Computers may be a recent invention, the idea of replacing continuous quantities by discrete approximations is quite old and lies at the heart and at the beginning of mathematics.

Calculating the circumference of a circle by approximating it by better and better fitting polygones for which the length is easily calculated, is an example from ancient times. In fact, just as the continuum hypothesis in the physical description of fluids, it could equally well be said that the continuous circle is an idealisation of a polygone-like construction. Other examples are Riemann sums for calculating integrals, the difference quotient for the derivative etc. And again, it can well be said that the mathematical definitions of integral and derivative are idealisations, limiting quantities, of the discrete concepts. As a final example, remember the representation of periodic functions on an interval by their Fourier series. Since there are infinitely many different base functions, the sine- and cosine functions, the set of periodic functions is called an infinite dimensional function space. *Infinite* to distinguish from a finite dimensional vector space for which a finite number of base vectors suffice to form each element by linear combinations. By considering a truncation of the series to some order, the infinite dimensional function space reduces (is "discretised") to a finite dimensional one which consists of vectors for which the elements are the finite number of Fourier coefficients. The other way around would be to start with finite sums of harmonic functions, and then consider linear combinations of more and more base functions, leading from finite dimensional spaces to an infinite dimensional one in the limit.

In the following, we will consider some of the basic questions that are related to the process of discretising. Starting with given quantities and equations of a continuous nature, we will address the question which properties a discretisation should have, and how these requirements are related to the actual purpose the discretisation should serve.

Approximation

Considering functions defined on some interval $I \subset R$, a function is defined once its value is specified at each point of I. This requires an infinity of data for each function. So, a nontrivial (linear) space of functions, say U, is infinitely dimensional. An important part of *approximation theory* deals with the question how, and how well, a given continuous function can be approximated if only a finite number of data are given. The simplest example is to define a "grid", i.e. a partition of the interval, to take the value of the function u at the grid points, and to specify a prescription how the approximation will look like in between. Linear, quadratic, cubic or higher order interpolation between grid points are well known examples. Another discretisation is with finite Fourier truncations mentioned above, i.e. representing a periodic function with only a finite sum of harmonic functions.

In each of these methods, a quantity h can be introduced, for instance the (maximal) mesh width of the grid, or the inverse of the number of Fourier modes. Such an approximation will be denoted by u_h and is an element from a finite dimensional subspace U_h of U. In approximation theory convergence and other properties are studied. Sensibly speaking, the approximation procedure should be such that the smaller h is taken, the better u_h will "look like" the original function u:

$$u_h \to u \text{ as } h \to 0. \tag{1}$$

Specifying the norm in which sense the convergence holds, gives information about the quality of the approximation.

Remark. Finite dimensional linear spaces are closed for the usual linear operations: addition and scalar multiplication. One of the major problems in numerical calculations, where one often has to deal with nonlinearities, is that this is not the case for multiplication! The product of two piecewise linear functions is not piecewise linear; the product of two finite Fourier truncations contains in general more Fourier modes than the order of truncation. All these properties are well known, especially when one has in mind the same properties for decimal point arithmetic. Indeed, when dealing with numbers with only one decimal products may have two decimals (e.g. $0.1 \times 0.1 = 0.01$, which would be considered to be zero when approximating numbers with only one decimal).

Numerical convergence

Now assume that we have some "continuous system". Symbolically it can be written like

$$C(u) = 0$$

where C is some operation applied to the function u. For instance, this equation can be an algebraic or differential equation. Let C_h be some "discretisation" of C, i.e. an operation defined on a space of approximations u_h, and consider the equation

$$C_h(u_h) = 0.$$

Then it is said, roughly speaking, that the discretisation procedure is *numerically convergent* if solutions of the discretised equation tend to the solution of the equation for vanishing h:

$$\text{if } C_h(u_h) = 0 \text{ and } h \to 0, \quad \text{then } u_h \to u \text{ with } C(u) = 0. \tag{2}$$

Just as is the case in approximation theory with condition (1), numerical convergence expressed by (2) is a natural requirement for any discretisation. Many investigations in this area concentrate on proving the numerical convergence for a given discretisation (using such concepts as numerical consistency and - stability). The norm(s) in which the convergence holds provides, just as for functions, a measure for the quality of the discretisation.

Model consistency

In approximation theory, when attention is restricted to functions with a specific property (such as the degree of differentiability, or a monotonicity or convexity property for instance), one may want to restrict to approximations with the same properties. The same could be asked when discretising equations: if for the given continuous system some property is known to hold (possibly only for a limited set of solutions), one would like to find numerically convergent discretisations which possess the same property. Although the formulation here is rather vague, the question is quite relevant, as shall be explained in the rest of this contribution. Clearly, what is required in order to achieve this aim, is some knowledge of the original system that is essential (and analytically manageable enough) to preserve under discretisation.

In this paper we will deal with dynamical systems, and show that properties such as (time-) asymptotic behaviour, conservation of energy, and the appearance of travelling waves can be inhereted by suitably constructed discretisations, at a fixed h.

We will refer to discretisations which retain such a specific property as *model consistent discretisations*, since such discretisations will "model" some essential property of the continuous equation. Compared to multi-purpose discretisation methods, the construction of model consistent discretisations requires additional analytical insights.

Furthermore, there is an essential difference with the concept of numerical convergence, which is a property of the discretisation for $h \to 0$. Model consistency, on the other hand, requires that a special property of the continuous equation is present in the discretisation at a *fixed h*.

2 Logistic equation

The *logistic differential equation* is the first order differential equation for a scalar function x of time $t : x = x(t)$ given by

$$\dot{x} = x(1 - x) \tag{3}$$

To give just one interpretation of this equation, when the variable x denotes the size of some population, and $x(t)$ is the size at time t, this equation describes the restricted growth of the population. The variables have been normalised in such a way that possible constants can be taken to be 1.

Given a certain *initial value*, i.e. the value of x at some initial time (which may be taken without restriction to be $t = 0$), say $x(0) = x_0$, with x_0 a given number, the solution of (3) can be written down explicitly (using the method of separation of variables). The qualitative behaviour of the solutions is depicted in Fig.1. (For the interpretation of x as the size of a population, negative values of x are not realistic and could be discarded.)

Now consider the most direct method to discretise (3). The unknown function $x(t)$ which is sought to satisfy the differential equation, is replaced by function values at gridpoints. For the first grid point we take $t = 0$, and next points a distance h apart. Then we look for a discretisation of (3), i.e. a procedure to calculate values $\hat{x}_n$ which are an approximation for $x(nh)$.

To that end, the simplest way is to replace the derivative in (3) by the (forward) difference quotient. The expression that arises is known as the *Euler discretisation*:

$$\hat{x}_{n+1} - \hat{x}_n = h\hat{x}_n(1 - \hat{x}_n). \tag{4}$$

This equation can be rewritten in a simpler way by introducing a normalised variable $y_n = a\hat{x}_n$ with $a = \frac{h}{1+h}$. Then the way how y_{n+1} follows from y_n can be written concisely: with $y = y_n$ and then writing y' for the next value:$y' = y_{n+1}$, the relation is given by

$$y' = \mu y(1 - y) \text{ with } \mu = 1 + h. \tag{5}$$

This map $y \to y'$ is known as the *logistic map*, and can be depicted by plotting y versus y' as in Fig.2.

The fixed points of this map are $y = 0$ and the more interesting one, which we will call $\bar{y}$, is given by $\bar{y} = \frac{\mu-1}{\mu} = \frac{h}{1+h}$. These fixed points $y = 0$ and $\bar{y}$ correspond to the equilibrium solutions $x(t) = 0$ and $x(t) = 1$ respectively of the logistic differential equation.

Applying the logistic map several times to a given initial condition, produces a sequence of points y_n, or $\hat{x}_n$ which should be approximants for $x(nh)$ with $x(t)$ the solution of (3). In between the points $\hat{x}_n$ we imagine the function to be approximated by linear interpolation. It is intuitively clear that the smaller the value of h, the better this piecewise linear function will approximate the solution of the continuous differential equation on compact time intervals. (A mathematical proof of this intuitive idea will be ommitted here.) This means that the Euler discretisation scheme is numerically convergent.

Note, however, that even for the relatively large value of h in Fig.2, and despite the difference with the trajectories depicted in Fig.1, the asymptotic behaviour is the same: the solution $\hat{x}_n$ tends to the stable equilibrium $x = 1$ for increasing n, or, equivalently, the sequence y_n converges to the nontrivial fixed point $\bar{y}$ of the discrete map (5).

Increasing further the value of h, a qualitative change in the asymptotic behaviour appears as h crosses the value 2. For $h > 2$, i.e. $\mu > 3$, the fixed point $\bar{y}$ is no longer stable: arbitrary points, even those starting arbitrarily close, are no longer attracted to $\bar{y}$. What actually happens, depends on the precise value of h: for $2 < h < 2.44$ there appears a socalled attracting period 2 solution of (5): the values $\hat{x}_n$ tend to switch between two fixed values around $x = 1$, see Fig.3. For $2.44 < h$ an attracting period 4 solution appears, and the period doubling process is repeated untill for $h > 2.57$ chaotic behaviour appears: the successive values $\hat{x}_n$ jump from values above 1 to values below 1 in a "random, chaotic way".

Remark. The logistic map serves as the prototype map in bifurcation theory to show the appearance of chaos through period doubling process. The bifurcation diagram is shown in Fig.4; interested readers are referred to the literature, e.g. Guckenheimer & Holmes (1983), Devaney (1986), Schuster (1987), and Broer e.a. (1991).

From these observations, we can make some valuable conclusions explicit. Firstly, the Euler discretisation method is numerically convergent. More than that, although the discretisation is rather poor initially for stepsizes h not close to 0, for each $h < 2$ the asymptotic behaviour of the solutions is well captured by the discretisation. Hence it can be said that for each fixed time step $h < 2$ the Euler discretisation is "model consistent" as far as the asymptotic behaviour is concerned. It may also be remarked that even in this rather simple case, a detailed study of the relation between the differential equation and its discretisation is rather involved. This should make one modest in expectations when dealing with more complicated situations.

Concerning the appearance of chaos it may be argued that this only appears for unrealisticcally large values of h, namely for $h > 2$. Indeed, for the isolated problem at hand this is true. But as we shall see later, in more complicated equations with several quantities, it may very well be the case that a time step which is small for one quantity is large for another one. Then a very good approximation of one quantity may be accompagnied by a (numerically enforced) chaotic behaviour of another one, which may influence the complete result. See also the Epilogue.

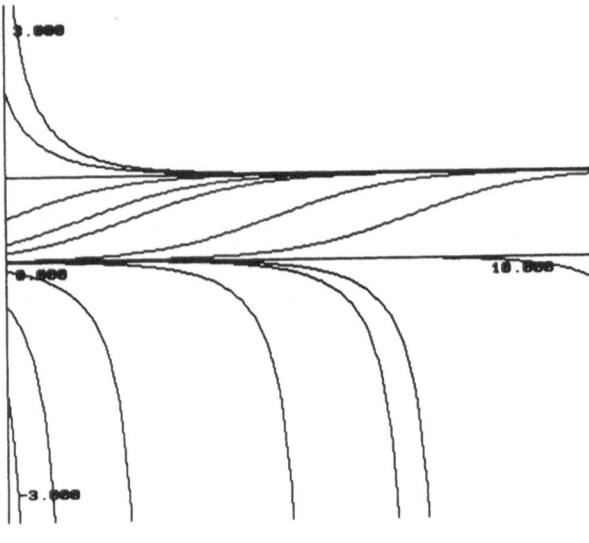

Figure 1:

Solution curves for the logistic d.e. $\dot{x} = x(1 - x)$, plotted x versus time t horizontally. There are two *equilibrium solutions:* $x(t) = 0$, and $x(t) = 1$ for all time. For initial data x_0 between 0 and 1, the solution exists for all time, and tends to 1 for $t \to \infty$ and to 0 for $t \to -\infty$. For initial value $x_0 > 1$ the solution exists for all positive times, and tends to 1 for $t \to \infty$. For $x_0 < 0$ the solution blows up to $-\infty$ for a finite positive time. Because the equilibrium 1 attracts all solutions starting sufficiently close, this equilibrium is called *stable.* The equilibrium solution 0 is *unstable.*

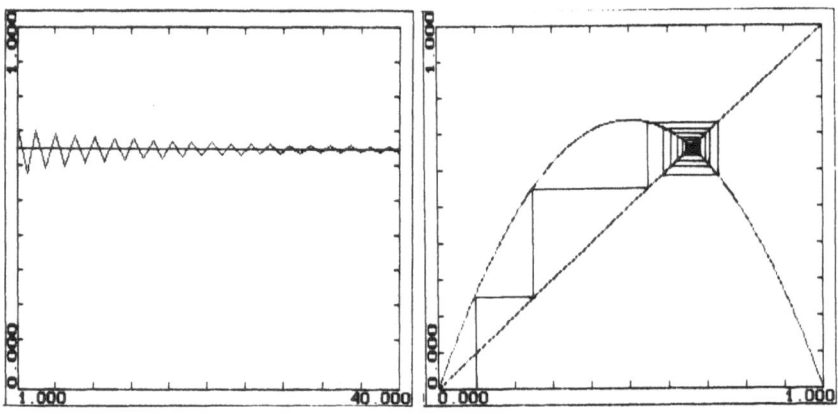

Figure 2:

In the left picture the piecewise linear function is drawn obtained from Eulers discretisation method, using $h = 1.95$, for 40 iterations. In the picture at the right, the same data are plotted in the normalised variable y as a map.

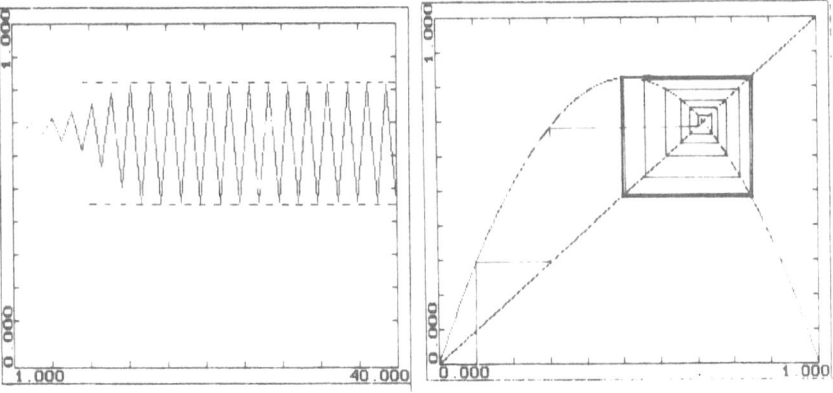

Figure 3:

As Fig.2, but now for $h = 2.3$, for which a fast convergence to a period two solution of the map is seen. The corresponding piecewise linear curve in x vs. t will not converge asymptotically to the equilibrium solution $x(t) = 1$ of the logistic differential equation.

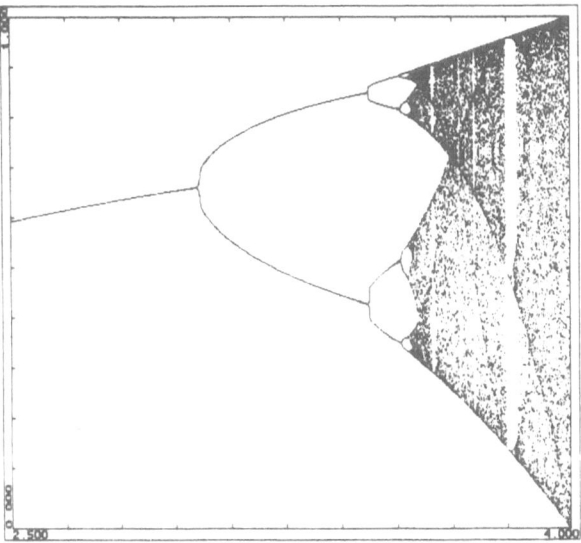

Figure 4:

The bifurcation diagram for the logistic map (5) with horizontally the bifurcation parameter μ. Note the period doubling process, starting at $\mu = 3$, after which the fixed point is no longer stable: for $\mu > 3$, corresponding to stepsize $h > 2$, the discretisation (5) has a different asymptotic behaviour then the continuous logistic differential equation. The wonderfull, complicated scaling processes in the chaotic area of this picture is explained in the relevant litterature.

3 Harmonic Oscillator

Consider another well known simple differential equation, now of second order, for
the scalar function $q(t)$:

$$\ddot{q} = -\omega^2 q. \tag{6}$$

This is the *harmonic oscillator*, which has as general solution the harmonic functions with frequency ω:

$$q(t) = A\sin(\omega t) + B\cos(\omega t).$$

The constants A and B can be adjusted to satisfy given initial data for q and its
first derivative $\dot{q}$.

Written as a system of two first order equations, (6) becomes

$$\begin{aligned} \dot{q} &= \omega p \\ \dot{p} &= -\omega q \end{aligned} \tag{7}$$

Introducing the two-vector $z = (q, p)$, this is more concisely written as

$$\dot{z} = \omega J z \tag{8}$$

where J is the matrix: $J = \begin{pmatrix} 0 & 1 \\ -1 & 0 \end{pmatrix}$. This matrix is skew-symmetric and
satisfies $J^{-1} = -J$, and is called a *symplectic* matrix. It is possible to scale down
the coefficient ω in (7) by introducing a new variable $\tau = \omega t$. Writing $\hat{z}(\tau) = z(t)$,
the equation becomes

$$\frac{d}{d\tau}\hat{z} = J\hat{z}. \tag{9}$$

For the rest of this section we will simply write z for $\hat{z}$, i.e. consider (8) with
$\omega = 1$.

When performing numerical calculations, this scaling should be kept in mind. In
fact, in view of the sinusoidal solution, it is clear that at least two time steps
are required per period to capture the least required information about such a
solution. Hence the maximum time step h should satisfy $\omega h \leq 2\pi/2$, i.e.

$$h \leq \frac{\pi}{\omega}. \tag{10}$$

For later reference we note that the equations above are a Hamiltonian system.
Indeed, introducing the *total energy* H as the Hamiltonian:

$$H(z) = \frac{1}{2}|z|^2 = \frac{1}{2}(q^2 + p^2), \tag{11}$$

the equations can be written like

$$\dot{z} = J\nabla H(z), \quad \text{equivalently} \quad \begin{aligned} \dot{q} &= \frac{\partial H}{\partial p} \\ \dot{p} &= -\frac{\partial H}{\partial q}. \end{aligned} \tag{12}$$

Equations of this form are called Hamilton equations for a system with Hamiltonian H. One particular property of such equations is that the total energy is conserved. Indeed, as a consequence of the skew symmetry of J, it holds for solutions of (12) :

$$\frac{d}{dt}H(z(t)) = z \cdot Jz = 0.$$

Fig.5 summarises the most important properties of the solutions of the system. The *phase portrait* is obtained by plotting $q(t)$ versus $p(t)$ upon eliminating the time t. The trajectories are circles centered around the equilibrium solution $q = 0$, $p = 0$. This equilibrium is called a *center*.

In every point in the phase plane, the direction field is tangent to the trajectory (circle) through the point. This immediately explains why the Euler discretisation method from the foregoing section will produce solutions that diverge away from circles. See Fig.6. This divergence is present for all values of the stepsize h, of course less prominent for small values of h.

When long time calculations are to be made for which the energy should be conserved, the choice for a higher order method such as Runge Kutta may seem to be obvious. Indeed, RK4 is a fourth order method and conserves the energy much better than the first order Euler scheme. However, RK4 is a scheme that is somewhat dissipative, so that despite the higher order accuracy, for large times the energy will decrease anyhow (see Fig.10).

There are several ways to retain energy conservation in calculations exactly. We will describe two essentially different approaches. The first one is of a numerical nature and consists of modifying a given discretisation method. The second one is more analytical and modifies the given differential equation in such a way that, (almost) independent of the discretisation method to be used, the calculations have the desired property of energy conservation.

3.1   Modification of the discretisation

When a given discretisation method can be studied in detail, the reason of the change in the energy can be investigated, and one may be able to improve energy conservation in a direct way. This can only be done for simple schemes. In this subsection we will demonstrate the idea to Euler's scheme considered in the preceding section.

Euler's scheme for $\dot{z} = Jz$ can be written as the map

$$z \rightarrow z' \text{ given by } z' = (I + hJ)z, \tag{13}$$

where I denotes the identity matrix and h is the stepsize. The energy E' after one step is easily found to be related to the previous energy E as:

$$E' = \tfrac{1}{2}|z'|^2 = \tfrac{1}{2}(1 + h^2)|z|^2 = (1 + h^2)E \tag{14}$$

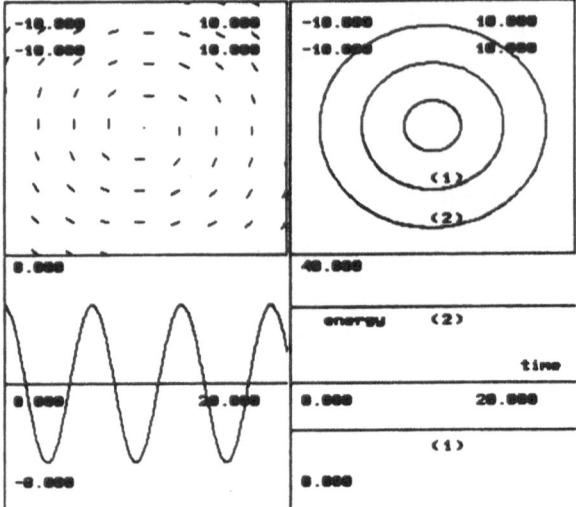

Figure 5:

The harmonic oscillator. UL: The upper left picture is a plot of the direction field. UR: Some phase curves are shown in the upper right picture, motivating the name 'center' for the equilibrium at the origin. LL: In the lower left picture the component q is plotted versus time, showing the harmonic behaviour. LR: The lower right picture shows the energy versus time, constant for the solutions shown in the phaseplane.

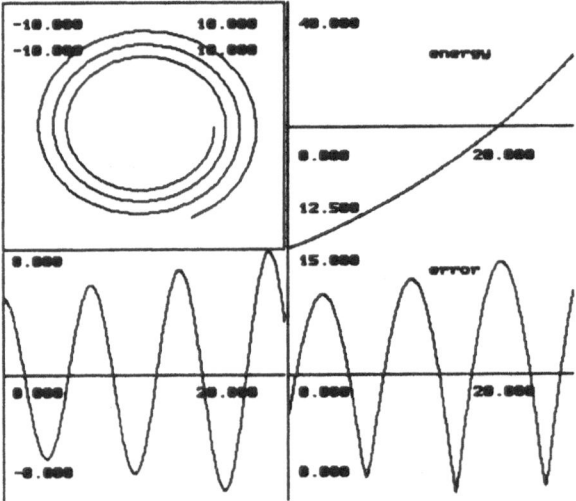

Figure 6:

Euler discretisation of the harmonic oscillator, with stepsize $h = 0.05$. UL: A trajectory diverges quickly, even for this small value of h. UR: The energy increases monotonically for the calculated solution. LL: The component q versus time shows an increase in amplitude. LR: The *error* as a function of time. The error is defined as the distance in the phase plane between the calculated solution and the exact solution. Except from the divergence (energy increase), this error is mainly due to *phase error*. See also the next figures.

This explains the monotone increase of the energy as depicted in Fig.5. To modify the scheme such that it conserves energy, artificial dissipation has to be introduced to compensate for the energy increase. This can easily be done by introducing a dissipation in the discretisation (13), the strenght of which is controlled by a parameter α, the value of which has to be determined suitably. For the map

$$z \to z' \text{ given by } z' = ((1 - \alpha h)I + hJ)z, \tag{15}$$

the energy change in one step is given by

$$E' = \gamma E \text{ with } \gamma = (1 - \alpha h)^2 + h^2. \tag{16}$$

Hence the scheme (15) is energy conserving for the choice

$$\alpha = \frac{1 - \sqrt{1 - h^2}}{h} \approx \tfrac{1}{2}h. \tag{17}$$

This energy conserving scheme can be interpreted in several ways. One way is to note that the modified discretisation (15) is in fact Eulers method with stepsize h for the modified differential equation

$$\dot{z} = Jz - \alpha z. \tag{18}$$

This differential equation is dissipative with an exponential decrease of the energy: $\frac{d}{dt}H(z) = -2\alpha H(z)$. So, this dissipation compensates for the energy increasing effect of the Euler scheme.

Remark. A second interpretation of (15) is more interesting. For the chosen value of α the discretisation scheme can be formulated, after some calculation, like

$$z' - z = \bar{h}J(\tfrac{1}{2}(z + z')) \tag{19}$$

where $\bar{h}$ is related to h by $h = \bar{h}/(1 + \bar{h}^2/4)$. This is a simple example of a *symplectic* discretisation scheme. Such schemes get much attention nowadays and are especially designed to retain an important property of classical Hamiltonian systems. This property is that the time evolution is a symplectic map (implying, for instance, that the Hamiltonian flow is volume preserving in the phase space). The last scheme has this property too since the defining map is symplectic, which means that $z' \cdot Jz' = z \cdot Jz$, as can easily be verified. Energy conservation, however, is not guaranteed for symplectic schemes. The conservation property in this particular example is a consequence of the additional property that the equations are linear.)

3.2   Modification of the differential equation

In this subsection we consider a completely different method to improve energy conservation in numerical schemes. Instead of looking into the properties of the schemes itself, we will modify the differential equation itself. The modification will be such that each discretisation will more succesfully conserve the energy for solutions that have the same initial energy (the discretisation will be different

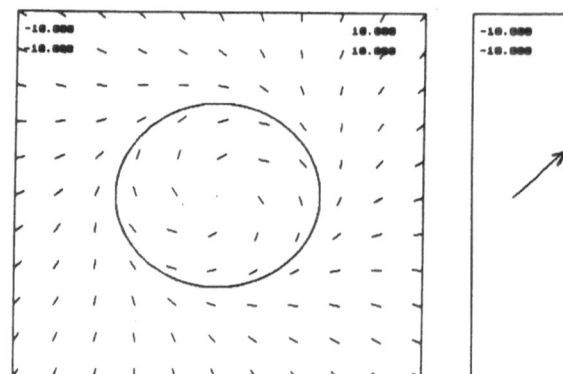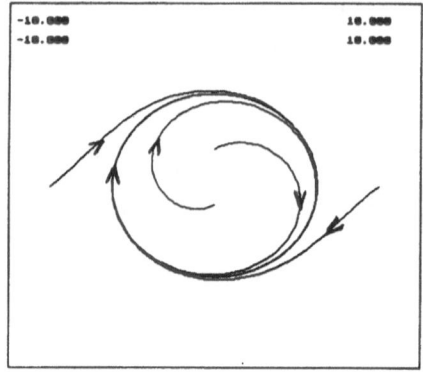

Figure 7:

For the modified equation (20) the direction field is shown in the left picture. Trajectories starting outside the level set E_0 are shown to be attracted to this set in the right picture.

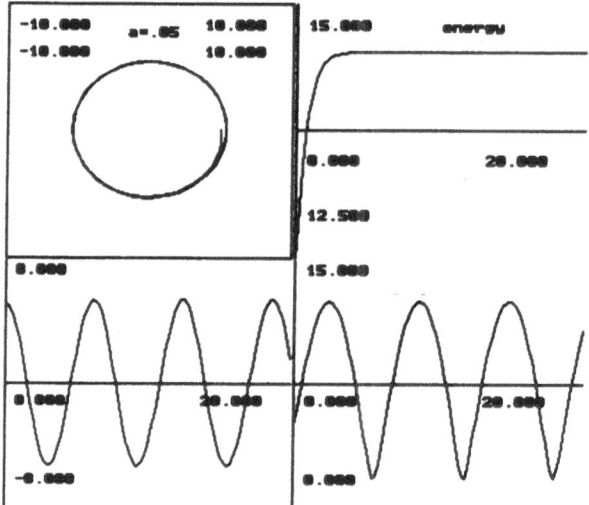

Figure 8:

Euler discretisation of the modified differential equation (20) for time step $h = 0.2$ and parameter $a = 0.05$. The initial data has energy $E_0 = 12.5$, and the asymptotic value is approximately 13.5, in agreement with (23). UL: trajectory of solution starting at E_0, LL: the $q-$ component vs time, UR: the evolution of the energy for the calculated solution, LR: the error of the solution.

for different values of the initial energy). For given initial data x_0 let E_0 be the corresponding value of the energy. For the analytical solution of (12) the energy is conserved $H(x(t)) = E_0$ for all time, but not for a solution from numerical calculations, as is seen from (14). Therefore we will modify the differential equation in such a way that the level set E_0 becomes attractive: trajectories that have deviated from this level set (by numerical errors) will be forced back to this level set. To that aim consider for some suitably chosen parameter α the modified equation:

$$\dot{z} = J\nabla H(z) - \alpha(H(z) - E_0)\nabla H(z), \tag{20}$$

explicitly

$$\dot{z} = Jz - \alpha(H(z) - E_0)z.$$

The additional term $-\alpha(H(z) - E_0)\nabla H(z)$ in the equation acts like a driving force with the desired properties. This can be seen by considering the evolution of the energy which is given by

$$\frac{d}{dt}H(z) = -\alpha(H(z) - E_0)|\nabla H(z)|^2. \tag{21}$$

For the case under consideration of a quadratic H this becomes the logistic equation for H:

$$\frac{d}{dt}\left(\frac{H(z)}{E_0}\right) = 2\alpha E_0 \left(\frac{H(z)}{E_0}\right)\left(1 - \frac{H(z)}{E_0}\right) \tag{22}$$

From this it follows that the analytical solutions of (20) are attracted to the level set E_0. In Fig.7 this property is displayed pictorially: the direction field clearly shows the presence of the modification (compare with Fig.5).

Now discretisations of the modified differential equation will preserve the energy much better. This is shown in Fig.8 for the Euler discretisation and in Fig.9 for the fourth order Runge Kutta method. Note that for this simple example the modification makes it possible to take much larger time steps than would be possible without the modification. The conservation of energy in large time calculations is shown in Fig.10.

Remark As we can see from the numerical calculations, the energy is well conserved, although the asymptotic value E_∞ differs from the initial value E_0 that had to be retained. This is due to the net result of two effects in these calculations: the dissipative effect from the modified equation (when $\alpha > 0$), and the property of the numerical scheme: slightly dissipative for RK4, and energy increasing for Euler.

To understand some of the results, consider once again the Euler method. A simple calculation shows that the energy changes per step as in (16) but now with a modified coefficient:

$$E' = \gamma E \text{ with } \gamma = (1 - \hat{\alpha}h)^2 + h^2,$$

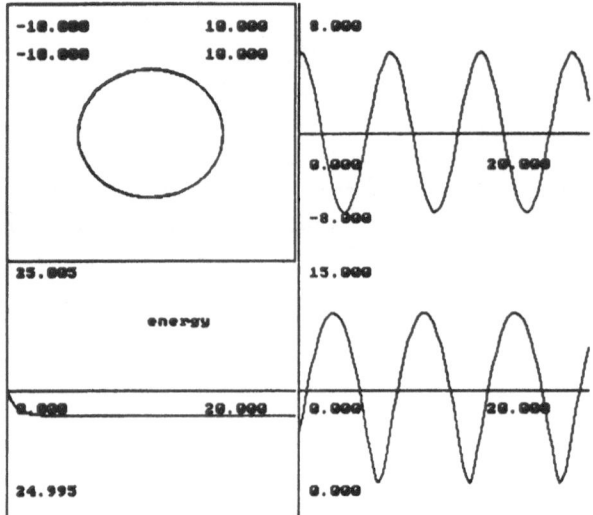

Figure 9:

Runge-Kutta 4-th order discretisation of the modified differential equation (20) for time step $h = 0.2$ and parameter $a = 0.03$. Same display as in Fig.8, but note the different scale along the energy axis (multiplied by 2) in the LL picture.

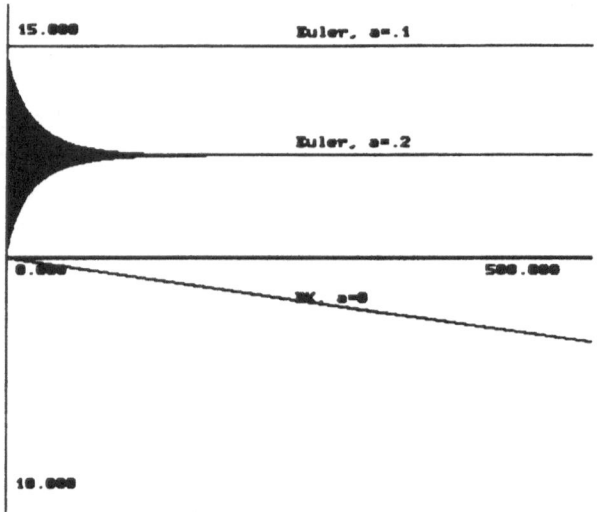

Figure 10:

Calculated energy during large time calculations with large time step $h = .04$ for Euler (upper) and Runge Kutta (lower). Note the different asymptotic values E_∞ for Euler depending on the choice of the parameter $a = \alpha$, agreeing with (23). For $\alpha >= .2$ Euler's scheme becomes unstable. For $\alpha = 0$ RK4 is seen to be dissipative. For $\alpha = .05$, calculations with RK4 are indistinguishable in this picture from the exact value at the axis.

with

$$\hat{\alpha} = \alpha(H(z) - E_0).$$

Hence, in the pictures the asymptotic energy follows with (17):

$$\alpha(E_\infty - E_0) \approx \tfrac{1}{2}h. \tag{23}$$

Remark. Despite the fact that we succeed to control the energy of the oscillation, at a value more or less close to the desired value, this has hardly any effect on the error of the calculated solution itself. Compare the error in Figs.6,8 and 9. The reason is that the error, defined as the distance in phase plane between the exact and the calculated solution, is mainly caused by an error in phase. So, even when we are able to keep the calculated solution well on the circle with the correct radius (determining the energy), the place at this circle at a certain time will be different from the exact position. When the calculated solution is "out of phase" with the exact one, the maximal error is found, being twice the radius. This phase error depends on the frequency, i.e. on the value ω in (8). When dealing with only one oscillator, this phase error can be suppressed by using a special discretisation that is based on knowledge of the exact solution. When two or more oscillators have to be calculated at the same time, as will be the case in the next section, this phase error cannot be suppressed when the frequencies are different.

4   Wave equations

In this section we will consider a much more difficult example than in the previous sections. Firstly, the equation will be a partial differential equation instead of an ordinary differential equation (a relation between a function of more than one variable and its partial derivatives). Secondly, different from the foregoing example, some of the equations to consider will be nonlinear, the importance of which will be explained. To start, however, we consider the simplest possible equations.

4.1   Linear first order wave equations

Let u be a scalar function of two variables, the time t and a spatial coordinate x : $u = u(x, t)$. Partial differentiation with respect to t wil be denoted by ∂_t, and partial differentiation with respect to x as ∂_x or with a subscript x as in u_x or u_{xx} for the first and second partial derivative respectively. Consider for a given constant c_0 the following equation:

$$\partial_t u = -c_0 \partial_x u \tag{24}$$

This partial differential equation is of first order in time, and we can write down the solutions explicitly: for arbitrary function f the function

$$u(x, t) = f(x - c_0 t) \tag{25}$$

satisfies (24). In fact, (25) is the unique solution of (24) which satisfies the *initial condition*

$$u(x,0) = f(x).$$

With the interpretation of u as the heigth of the free surface of a layer of water which extends in the x- direction, measured from some horizontal reference level, it will be clear why equation (24) is called a *wave equation*, and the solution (25) a *uniformly travelling wave*, see Fig.11.

In the x vs. t -plane the solution can be found from the observation that on lines $x - c_0 t$ = constant the function u satisfying (24) is constant. Stated differently, on the socalled *characteristics* the partial differential equation reduces to an ordinary differential equation:

$$\frac{d}{dt} u(x(t), t) = 0 \text{ on } \frac{d}{dt} x(t) = c_0.$$

For a numerical integration of the partial differential equation (and more difficult ones for which such characteristics can be found), this property can be used advantageously. Indeed, a simple *finite difference method* can be constructed by taking Euler discretisations in both the x and t-variable, for t the forward-Euler method with step Δt, and for x the backward difference with step Δx. Then the information about the value of u can be transported without distortion provided the quotient of the stepsizes is chosen appropriate. This is shown in Fig.12. Note that if the quotient $\frac{\Delta x}{\Delta t}$ is chosen different from the speed c_0, this finite difference scheme will deform the solution in an unacceptable way, as in Fig.13, or the scheme will be unstable.

The following wave equation shares some, but not all, properties of (24):

$$\partial_t u = -c_0 \partial_x (u + \alpha u_{xx}). \tag{26}$$

Special solutions, socalled *monochromatic solutions*, can be found of the following form. For each value of the *wave number k*, the (real and imaginairy part of the) function

$$\varphi_k(x, t) = exp(i(kx - \omega t)) \tag{27}$$

is a solution provided the *frequency ω* is related to k according to

$$\omega = \omega(k) = c_0(k - \alpha k^3). \tag{28}$$

Clearly, each monochromatic solution is a travelling wave solution, periodic in time and in space, propagating with velocity

$$v_{phase}(k) = \frac{\omega(k)}{k},$$

the socalled phase velocity. For $\alpha \neq 0$ this propagation speed depends on the wave number. This has the effect that a superposition of two monochromatic solutions

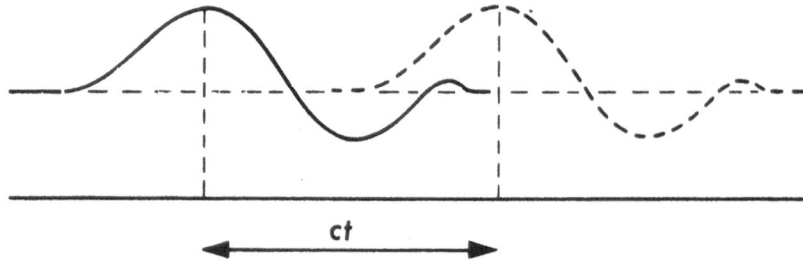

Figure 11:

With f the wave shape at some initial time $t = 0$, the corresponding solution $u(x, t)$ of (24) denotes the wave shape at later times. This solution $u(x, t) = f(x - c_0 t)$ is a *uniformly travelling wave*: the wave shape f is translated undistorted in shape with constant speed c_0 (in the positive x-direction if c_0 is positive).

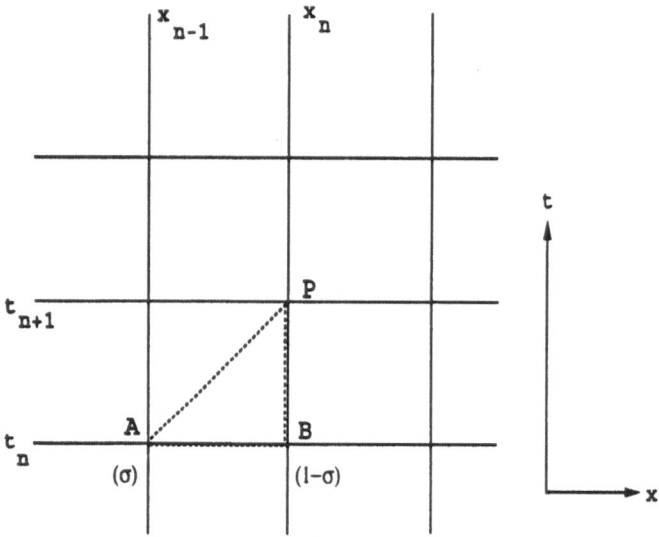

Figure 12:

Simple finite difference method for (27); forward Euler in time (step Δt) and backward Euler in space (step Δx). On a grid in the x, t-plane determined by the stepsizes, the value at a point P is determined from the values at A and B at a previous time. From $\frac{u_P - u_B}{\Delta t} = -c_0 \frac{u_B - u_A}{\Delta x}$ the weight of these contributiones follow as $u_P = (1 - \sigma)u_B + \sigma u_A$, with $\sigma = c_0 \frac{\Delta t}{\Delta x}$. For $0 < \sigma < 1$ this is a convex combination (linear interpolation) giving rise to a dissipative scheme, while for $\sigma > 1, u_P$ is obtained from an extrapolation, leading to an unstable scheme. Only for $\sigma = 1$ the correct value at P is found from that at A.

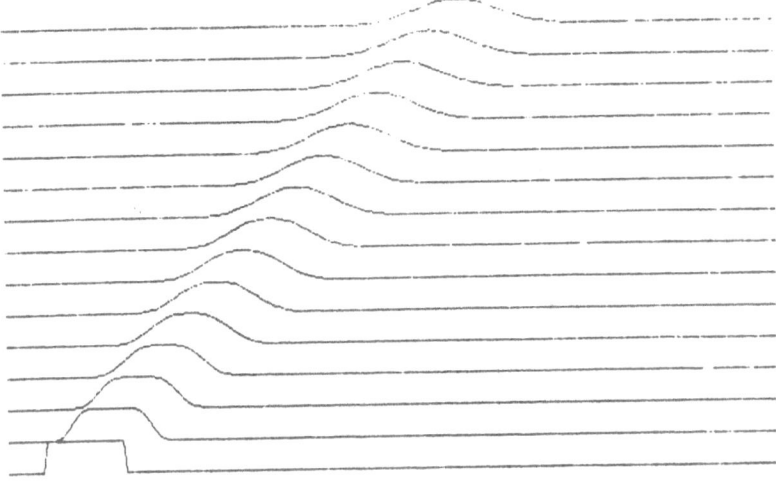

Figure 13:

Distortion from finite difference method for the simple equation (24). Shown are the calculated wave profiles at different times. The bottom line represents the initial profile: a block function. The line above that is the evolution after 20 time steps, and so on for 400 time steps. For this calculation $\sigma = 0.35$ (see Fig.12), and the dissipation of the block function is clearly visible.

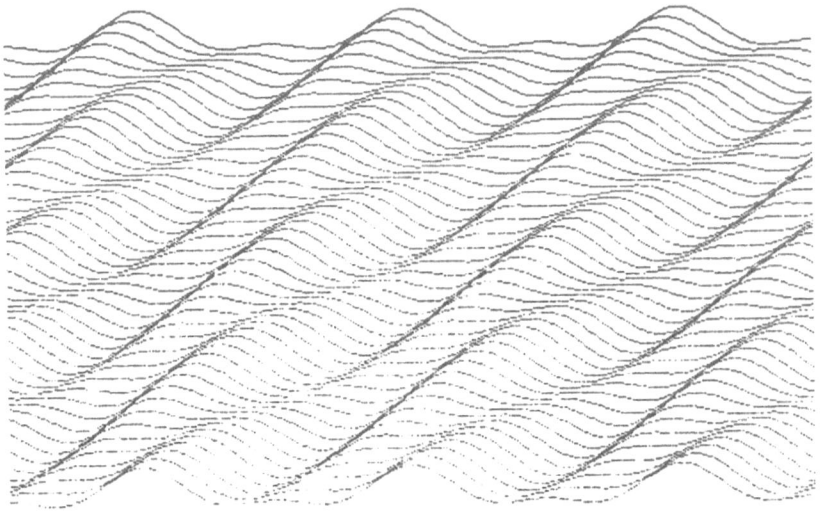

Figure 14:

Dispersion. Shown is the linear superposition of two waves in a frame of reference moving with the velocity of the largest wave. The smaller wave has relative amplitude $\frac{1}{3}$ and a three times larger velocity.

(two "modes") with different k will not behave as a uniformly travelling wave: the two modes will translate with different velocity and the total wave profile is the sum of the two shapes which translate with different speed. This effect is called *dispersion*, and the relation (28) is the dispersion relation. Equations such as (26) with this property are therefore called dispersive wave equations. See Fig.14.

The monochromatic solutions form a basis with which the general solution of (26) can be found using Fourier theory as follows. If an initial wave form is a given function f, either decaying sufficiently fast at infinity, or being a periodic function with, say, period 2π, its Fourier transform can be determined:

$$f(x) = \int_{-\infty}^{+\infty} \hat{f}(k)exp(ikx)dk, \text{ respectively } f(x) = \sum_{k=-\infty}^{+\infty} \hat{f}(k)exp(ikx).$$

Then the solution of (26) is the corresponding superpostion of monochromatic modes with the same spectral function $\hat{f}(k)$:

$$u(x,t) = \int_{-\infty}^{+\infty} \hat{f}(k)\varphi_k(x,t)dk, \text{ respectively } u(x,t) = \sum_{k=-\infty}^{+\infty} \hat{f}(k)\varphi_k(x,t).$$

So the analytic solution of the initial value problem for (26) can be written down explicitly using Fourier theory.

The numerical calculation of (26) is much more difficult than that of (24) when finite difference schemes are considered, since for $\alpha \neq 0$ no characteristics can be used. Replacing the spatial derivative with some difference quotient, either central-, backward - or forward difference, errors are introduced which will deform the wave shapes from the analytic solution.

When dealing with the periodic case, i.e. restricting to functions which are periodic (with period 2π for simplicity), an alternative way is to approximate the solution at each time by a truncated Fourier series, with Fourier coefficients depending on time and to be determined. Writing (using real quantities again, instead of the complex coefficients above) for functions with mean value zero

$$u(x,t) = \sum_{k=1}^{N} a_k(t)\cos(kx) + b_k(t)\sin(kx), \tag{29}$$

substitution into equation (26) leads to

$$\sum_{k=1}^{N} \left((\dot{a}_k - \omega(k)b_k)\cos(kx) + (\dot{b}_k + \omega(k)a_k)\sin(kx) \right) = 0.$$

The dynamic equations for the coefficients follow easily: for $1 \leq k \leq N$

$$\begin{aligned} \dot{a}_k &= \omega(k)b_k \\ \dot{b}_k &= -\omega(k)a_k. \end{aligned} \tag{30}$$

This shows that the partial differential equation has been reduced to a set of ordinary differential equation for the coefficients, and we are left with a generalisation of the problem considered in the foregoing section. Indeed, for each k the equation in a_k and b_k is nothing but that of a harmonic oscillator, with frequency $\omega(k)$, and the time integration can be performed as discussed previously.

Remark Note, however, that each mode oscillates with its own frequency $\omega(k)$. Hence, for a reliable numerical calculation of the k-th mode the time step h_k should satisfy according to (10): $h_k < \pi \cdot 1/\omega(k)$. Taking $N = 3$ as the highest mode, for which $\omega = -24$ has the largest absolute value, the condition (10) gives for the maximal timestep h

$$h < \pi \cdot 1/\omega_{max} \approx .13 \quad \text{for} \quad N = 3. \tag{31}$$

The results of Fig.16,17 and 18 indicate that (for the more difficult problem treated there) $h = .05$ is not sufficient, but that $h = .005$ (and in fact also $h = .01$) gives satisfactory results.

Hamiltonian formulation

Introducing the $2N$-vector $z = (a_1, b_1; ....; a_N, b_N)$, the equations can be written more concisely in various ways. For reasons to become clear shortly, we rewrite them as:

$$\dot{z} = -c_0 D \nabla H(z). \tag{32}$$

Here D is the 2Nx2N-block-diagonal matrix corresponding to the differential operator ∂_x for the truncated functions:

$$D = \text{diag}\,(J, 2 \cdot J, ..., N \cdot J), \quad \text{with } J = \begin{pmatrix} 0 & 1 \\ -1 & 0 \end{pmatrix}. \tag{33}$$

The function H is given by

$$H(z) = H((a_1, b_1; ....; a_N, b_N)) == \sum_{k=1}^{N} \tfrac{1}{2}(1 - \alpha k^2)(a_k^2 + b_k^2). \tag{34}$$

The point we want to make in writing the equations as in (32), is that this set of equations is of the form of a Hamiltonian system, akin to (12). Now the symplectic matrix J is replaced by $c_0 D$, which, although not symplectic anymore, is skew-symmetric: $D^* = -D$. This suffices to call (32) a *generalised Hamiltonian system*. More precisely, the equation is of the form of a *Poisson system*, a notion that does not have to be explained here in detail. The most appealing fact is that for equation (32) the Hamiltonian H is conserved, just as for classical Hamiltonian systems, as a consequence of the skew-symmetry of D:

$$\frac{d}{dt}H(z) = \nabla H(z) \cdot \partial_t z = -c_0 \nabla H(z) \cdot D \nabla H(z) = 0. \tag{35}$$

Since in this example H is just the sum of the energies of all harmonic oscillators, H is the *total energy* of the system, and (35) expresses that this total energy remains constant during the evolution.

In the next subsection it is shown that the same holds true for the more difficult case of non-linear equtions, for which the modes are no longer uncoupled as in (30), but are coupled through non-linear terms.

It may also be remarked here that the appearance of the Hamiltonian structure in the equations for the Fourier coefficients is not accidental. In fact, this structure is inhereted from the continuous partial differential equation (26), which is also in the form of a generalised Hamiltonian system. The operator $c_0\partial_x$ is a skew-symmetric operator (using the usual L_2-innerproduct for periodic functions), and the expression between brackets $u + \alpha u_{xx}$ can be seen as the "derivative" (the variational derivative in this case) of the functional which is the total energy $\bar{H}(u)$ of the continuous system. This energy is now a functional, given by

$$\bar{H}(u) = \int_{-\pi}^{\pi} (\tfrac{1}{2}u^2 - \tfrac{1}{2}\alpha u_x^2)dx, \tag{36}$$

and is also constant in time for solutions of (26) (which can be verified directly). There is no need to go any further in all these details here. But just note the optical resemblance of the equation (26) with the Hamiltonian form (32) when we write the variational derivative of $\bar{H}$ as $\delta\bar{H}(u)$ [1]:

$$\partial_t = -c_0\partial_x\delta\bar{H}(u). \tag{37}$$

Also, note that inserting the Fourier truncation (29) of u into the expression for $\bar{H}(u)$, performing the integrations (here, using Parcevals identity), the expression (34) results for H: H is nothing but the truncation of the energy $\bar{H}$ of the continuous system:

$$H(z) = H(a_1, ..., a_N, b_1, ..., b_N) = \bar{H}\left(\sum_{k=1}^{N} a_k(t)\cos(kx) + b_k(t)\sin(kx)\right). \tag{38}$$

Remark: In general the Hamiltonian structure will be destroyed when using finite difference schemes for (26), unless very special precautions are taken. For one thing this means that the energy will not be conserved. The Fourier truncation,

---

[1] For functionals defined on the set of periodic functions, like $\bar{H}$, the *variational derivative* is the generalization of the gradient of a function of a finite number of variables. In fact, it is defined as that function $\delta\bar{H}(u)$ such that for each other periodic function η it holds that

$$\int_{-\pi}^{\pi} \delta\bar{H}(u), \eta dx = \frac{d}{d\epsilon}\bar{H}(u + \epsilon\eta)|_{\epsilon=0}.$$

For the given expression for $\bar{H}$ it holds $\bar{H}(u + \epsilon\eta) = \bar{H}(u) + \epsilon\int_{-\pi}^{\pi}(u\eta - \alpha u_x\eta_x)dx + O(\epsilon^2)$. Integrating by parts the order ϵ term gives $\int_{-\pi}^{\pi}(u\eta - \alpha u_x\eta_x)dx = \int_{-\pi}^{\pi}(u + \alpha u_{xx})\eta_x \, dx$ which shows that $\delta\bar{H}(u) = u + \alpha u_{xx}$.

however, has retained the Hamiltonian structure of the original continuous system: the discretisation correctly *models* the basic structure, and hence the conservation property, of the original problem.

4.2 Nonlinear equation

For the linear equation considered above, each Fourier mode behaves independently from the other modes: the equations for a_k and b_k only contains these coefficients, no coefficients a_m or b_m with $m \neq k$. This is a consequence of the linearity of the original equation. When dealing with nonlinear equations, this does not hold any longer: the various Fourier modes will interact, causing a much richer (more difficult) dynamical behaviour.

As an example, we will consider the **Korteweg-de Vries equation**. This partial differential equation describes many physical phenomena in which dispersive and nonlinear effects are present in the same order of magnitude. For instance, the equation was derived originally to describe the evolution of surface waves on a layer of fluid, in the approximation of waves running (mainly) in one direction only. [2]

The equation is given by

$$\partial_t u = -c_0 \partial_x \left(u + \alpha u_{xx} + 3\beta u^2 \right), \tag{39}$$

or in a normalised form, which will be used in the following, by [3]

$$\partial_t u = -\partial_x \left(u_{xx} + 3u^2 \right). \tag{40}$$

This equation has several very special properties and has given rise to very far-going new developments in the theory of partial differential equations (complete integrability). At this place we will only touch briefly on some of these (less far reaching) properties.

The first observation is that, just as the linear equation (26), this equation has a Hamiltonian structure, i.e. is of the form of equation (37). Now the energy $\bar{H}(u)$ contains a non-quadratic term. For (40) the energy is given by

$$\bar{H}(u) = \int_{-\pi}^{\pi} \left(\tfrac{1}{2} u_x^2 + u^3 \right) dx. \tag{41}$$

The variational derivative of $\bar{H}(u)$ turns out to produce precisely the term in brackets in (40): $\delta \bar{H}(u) = u_{xx} + 3u^2$, and so the equation is indeed of the form

---

[2] The validity of this equation is restricted to "rather low, rather long" waves, for which dispersive and nonlinear terms are of comparable order, but small compared to the basic equation (24). This means that in reality the coefficients α and β in (39) should be small.

[3] When transforming to a frame of reference moving with velocity c_0, the equation (39) becomes: $\partial_t u = -c_0 \partial_x \left(\alpha u_{xx} + 3\beta u^2 \right)$. Then, by scaling the independent variables x and t and the variable u in an appropriate way, it is possible to arrive at the equation (40) in the scaled variables.

(37) of a Hamiltonian system.

This may lead immediately to the idea how consistent discretisations for (39) can be found using Fourier truncation as before. Indeed, based on this Hamiltonian structure, we could exploit the observation made for the linear equation. That is, substitution of the truncation for u in $\bar{H}(u)$ will produce a function of the Fourier coefficients, the truncated energy as in (38). Explicitly, after some tedious calculations (involving the integration of the product of combinations of three sine and cosine functions from the cubic term in the Hamiltonian), there results

$$H(z) = H_2(z) + \beta H_3(z), \tag{42}$$

where H_2 is the quadratic part as in (34), and H_3 is from the cubic terms.

The Hamiltonian being determined in this way, we take as discretisation of (39) simply the Hamiltonian equations given by (32). For each k they are of the form

$$\begin{aligned} \dot{a}_k &= -k\frac{\partial H}{\partial b_k} \\ \dot{b}_k &= k\frac{\partial H}{\partial b_k} \end{aligned} \tag{43}$$

For example, with $N = 3$, the Hamiltonian reads

$$H_3(z) = -1.5a_1^2 a_2 + 1.5b_1^2 a_2 - 3(a_1(b_2 b_1 + a_3 a_2 + b_2 b_3) + (a_2 b_3 - b_2 a_3)b_1) \tag{44}$$

and the corresponding equations are:

$$\begin{aligned} \dot{a}_1 &= -b_1 - 3(a_3 b_2 + a_2 b_1 - a_1 b_2 - a_2 b_3) \\ \dot{b}_1 &= a_1 - 3(a_3 a_2 + a_2 a_1 + b_1 b_2 + b_2 b_3) \\ \dot{a}_2 &= -8b_2 - 6(a_3 b_1 - a_1 b_1 - a_1 b_3) \\ \dot{b}_2 &= 8a_2 + 6(-a_3 a_1 - 0.5a_1^2 + 0.5b_1^2 - b_1 b_3) \\ \dot{a}_3 &= -27b_3 + 9(a_2 b_1 + a_1 b_2) \\ \dot{b}_3 &= 27a_3 + 9(-a_2 a_1 + b_1 b_2). \end{aligned} \tag{45}$$

Equations like those for the coefficients of the Fourier modes are usually referred to as mode equations. They can be interpreted to describe three *coupled nonlinear oscillators*, the coupling in this case being a consequence of the nonlinearity. Clearly, higher order Fourier truncations will lead to equally more mode equations, more coupled oscillators, all in interaction with (most of) the others.

Mode interaction

The equations (45) describe the three-mode interaction: due to the nonlinear terms, which result from H_3 in the Hamiltonian, the modes are coupled. For instance, the change in a_1 is determined not only by b_1 but also by the components of the other modes a_2, a_3, b_2, b_3 as is seen from the equations (45). See Fig.15.
It will be observed that the interaction process is quite complicated. Many small scale features are visible in the trajectory of the first mode, more clearly visible in

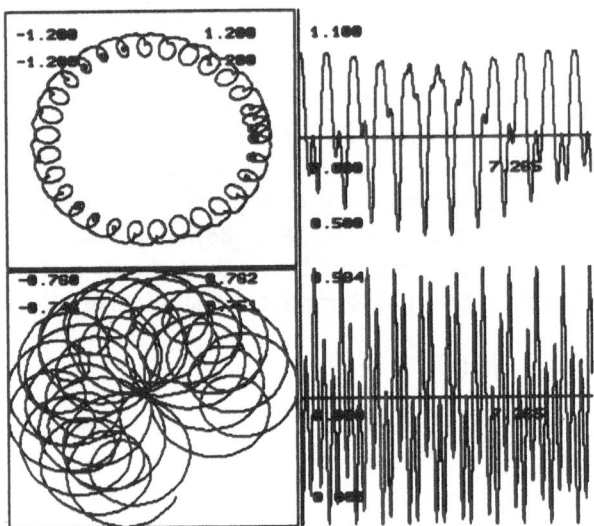

Figure 15:

Mode interaction for the 3-mode system (44). Starting with an initial condition for which the second mode is zero at $t = 0$, this mode is excited because of the nonlinear terms in (44). For initial data $(a_1, b_1; a_2, b_2; a_3, b_3) = (1, 0; 0, 0; 1, 0)$ the figure shows: UL: The evolution of the first mode in the a_1, b_1 plane of the six-dimensional phase space (for about a three times longer time period than the other pictures in this figure). UR: The evolution of the strength of the first mode $\sqrt{a_1^2 + b_1^2}$, which is not constant because of interaction with the other modes. LL: The evolution of the second mode (starting at zero) in the a_2, b_2 plane of the phase space. LR: The evolution of the strength of the second mode $\sqrt{a_2^2 + b_2^2}$, showing the excitation from and interaction with the other modes.

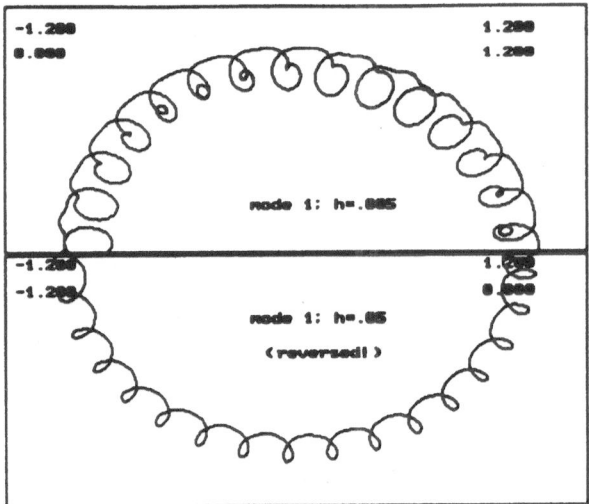

Figure 16:

Mode interaction for the 3-mode system (44). Starting with the same initial data as in Fig.15, the trajectory of the first mode is shown, using two different timesteps. In the upper picture the timestep is $h = .005$, the same as used to produce Fig.15 and 17. The lower picture has been calculated, for the same initial condition, with timestep $h = .05$, the same as used to produce Fig.18. Then the trajectory is transferred in the *opposite* direction! Moreover, the fine scale structure showing the presence of another time scale in the upper picture, is lost in the lower picture. The dependence on the timestep is drastically shown in this picture.

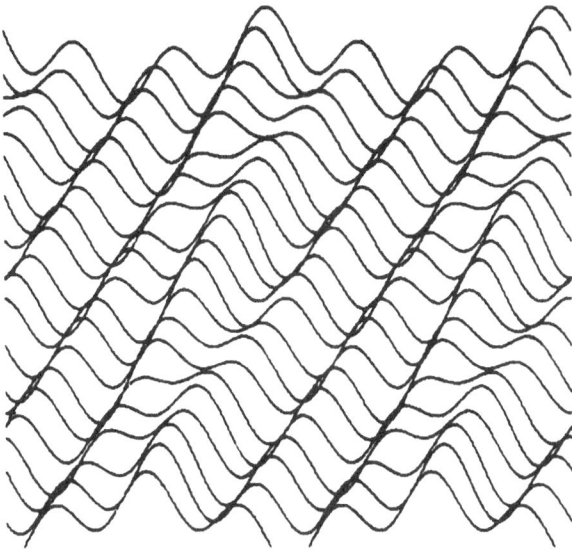

Figure 17:

Evolution of the wave profile $u(x,0) = \cos x + \cos 3x$, corresponding to the initial data $(a_1, b_1; a_2, b_2; a_3, b_3) = (1, 0; 0, 0; 1, 0)$ of Figs.15 and 16. The horizontal axis has length two times the spatial period: $2 \times 2\pi$. Shown are the profiles at different times: starting at $t = 0$, the next profiles are at $t = \frac{1}{2}k$, with $1 \leq k \leq 14$, each one shifted vertically upwards over a fixed amount. The timestep $h = .005$ corresponds to the time step used to produce Fig.15 and the upper part of Fig.16.

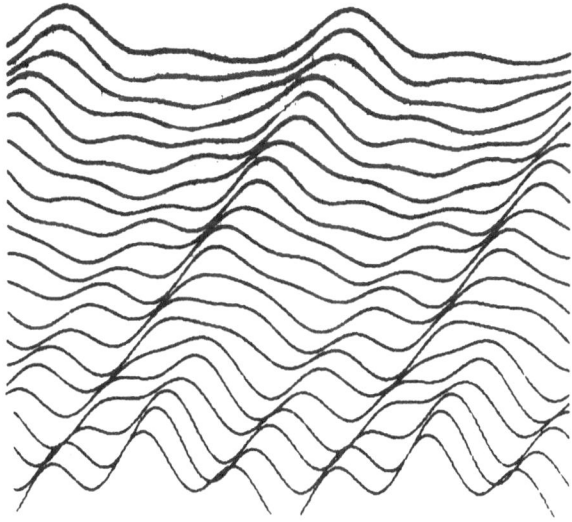

Figure 18:

Evolution of the same wave profile as in Fig.17, but now calculated with timestep $h = .05$, corresponding to the timestep used to produce the lower part of Fig.16. Observe that all small scale features are quickly fading away, resulting in a profile that differs essentially from the one calculated in Fig.17.

the upper part of Fig.16. The corresponding evolution of the wave profile is given
by the three term Fourier approximation (29) and is shown in Fig.17.

Comparing the lower part of Fig.16 with the upper part, and Fig.18 with Fig.17,
shows that the calculated results depend very much on the timestep. Even for
this simple three-mode model, the timestep has to be taken less than .01 in order
to obtain results that capture the fine scale structures. When calculating real life
problems, such as for instance atmospheric models for wheather forecasts, or flows
around wings, thousands and even millions of such coupled equations have to be
calculated. The relatively simple example considered here illuminates some of the
problems that are encountered there.

Travelling waves

Above we solved an initial value problem: given the profile at $t = 0$ we calculated
the profile at later times. Now we will deal with a completely different problem.
We will address the question whether there exists a *travelling wave*: a specific wave
profile that is translated at constant speed undeformed in shape.

For a nonlinear equation as the one under consideration, that will be an exceptional
solution, in contrast to the simple linear equations (24) for which each solution
is of this form, or the linear dispersive equation (26) for which monochromatic
solutions are of this kind.

We will investigate some aspects, both for the KdV equation itself and its N−mode
truncation. Upon substitution in (39) the general form of a travelling wave, say a
wave with profile $\hat{u}$ and velocity λ:

$$u(x, t) = \hat{u}(x - \lambda t), \tag{46}$$

there results an *ordinary* differential equation for the profile function $\hat{u}$ in which
the velocity λ enters as a parameter. It can be shown that this equation admits
(for appropriate values of the constants) 2π-periodic solutions. Since these solutions can be written down "eplicitly" using the elliptic function that is known as
the cnoïdal function, these waves are called *cnoïdal waves*.

Specifically, the equation for the function $\hat{u}(x)$ reads:

$$\lambda \partial_x \hat{u} = \partial_x \left(\hat{u}_{xx} + 3\hat{u}^2 \right)$$

and can be integrated once, introducing a constant of integration σ:

$$\lambda \hat{u} = \left(\hat{u}_{xx} + 3\hat{u}^2 \right) + \sigma. \tag{47}$$

This second order equation can be studied thoroughly. For instance, it can be brought
to a Hamiltonian form again (with x replacing the time variable). Consequently, its
Hamiltonian is constant (independent of x). This can also be found in a more direct way
by multiplying equation (47) by u_x and integrating again with respect to x. The result
is a first order equation for u of the form

$$u_x^2 = Q(u),$$

where $Q(u)$ is some cubic polynomial, depending on several constants, in particular on λ. Looking for a solution that is symmetric around $x = 0$ (without restriction of generality), this equation can be solved by separation as follows. If u_0 and u_π denote the value at $x = 0$ and $x = \pi$ respectively (the value at the crest and in the valley), which are chosen such that $Q(u) > 0$ for $u_\pi < u < u_0$, $Q(u_0) = Q(u_\pi) = 0$, then

$$\int_{u(x)}^{u_0} \frac{du}{\sqrt{Q(u)}} = x$$

describes implicitly the profile for $0 < x < \pi$. Periodicity requires that the coefficients are chosen such that $\int_{u_\pi}^{u_0} \frac{du}{\sqrt{Q(u)}} = \pi$. This still leaves a free parameter to adjust to the mean value of u, which can also be normalised to 0. See for more details e.g. Whitham, 1976.

Now that we know that the exact continuous equation has travelling waves, one could hope that its discretisation has such solutions too. In general that cannot be expected and the next statement is exceptional.

Proposition. *For each N, the N-mode Fourier truncation of the KdV-equation has exact travelling waves (2π-periodic in space).*

We will first present some "numerical evidence" for this statement, for $N = 3$, and then give the analytical arguments of the proof.

Numerical evidence.
It is *not* possible to find a periodic solution numerically by trial and error. We will show that the following set of data provide the desired travelling wave when $N = 3$. How these data are determined cannot be understood without referring to the analytic proof to follow. For the moment we just accept them as given. These data -for the resulting travelling wave velocity λ, and the Fourier coefficients- are as follows (the meaning and value of I will be explained later):

$$\text{for } I = 2\pi, \quad \begin{array}{rcl} \lambda & = & -3.65497558464, \\ \bar{a}_1 & = & 1.69505557741, \\ \bar{a}_2 & = & 0.97285831670, \\ \bar{a}_3 & = & 0.42465666678, \end{array} \quad \text{and} \quad \bar{b}_1 = \bar{b}_2 = \bar{b}_3 = 0 \qquad (48)$$

The Fourier coefficients are used as initial data for (45) and the resulting dynamics is calculated numerically with a small timestep. In the next figures this solution is depicted and interpreted in various ways. In Fig.19 the evolution of each of the three modes is shown in the phase space. For the specific initial value each mode behaves as an oscillator exhibiting a pure harmonic oscillation, without coupling with the other modes. This is shown in a different way in Fig.20.

The spatial wave profile corresponding to the Fourier coefficients of the special initial data is shown in the left picture of Fig.21. The dynamic evolution of the Fourier coefficients gives a profile that is shown in Fig.22. This picture gives optical evidence that the wave travels undisturbed in shape, at constant velocity, a

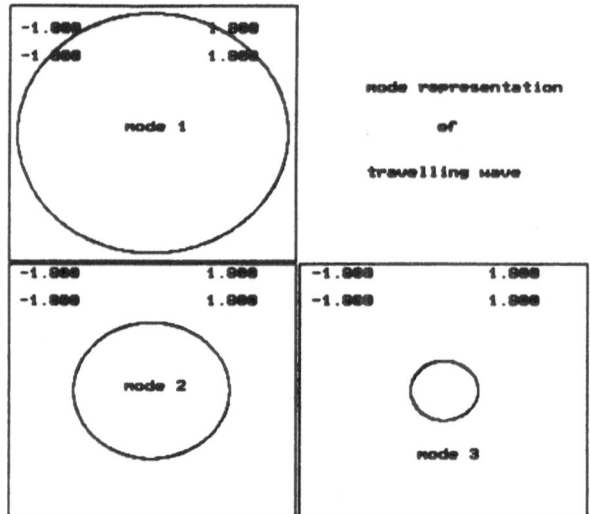

Figure 19:

For initial data given by (47), each mode behaves like a harmonic oscillator, with no visible sign of interaction. The time step used is $h = .01$.

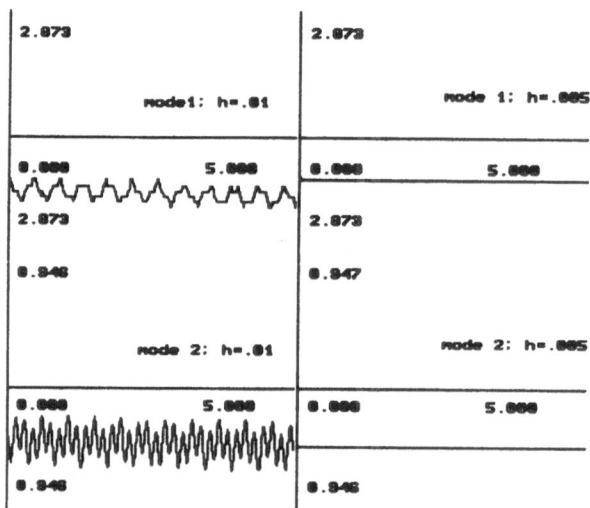

Figure 20:

A more precise representation shows that the amplitudes of the modes actually do vary a little bit (in the fifth digit), as is seen from the UL and LL picture for this timestep $h = .01$ (scales are: 2.87321-2.87322 for the first mode, and .94645-.94646 for the second mode). The pictures at the right show again these modes, now calculated for the same initial data but with a smaller timestep: $h = .005$. Comparison of the pictures at the left and at the right indicates that the variation in mode amplitude is only caused by numerical errors.

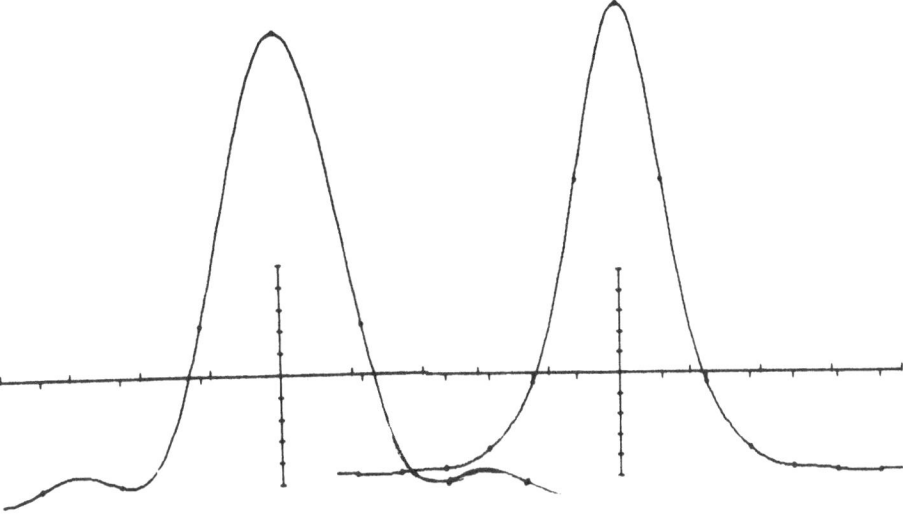

Figure 21:

The wave profile obtained with the N-mode Fourier truncation, using as coeeficients the data obtained from solving the constrained optimisation problem. In the picture at the left, calculated with $N = 3$, the low number of modes produce a wavy profile in the valley, which is not expected in reality. This wavy part disappears almost completely when more modes are taken into account: for $N = 6$ the profile, found by solving the optimisation problem with $N = 6$, is given in the right part of the picture.

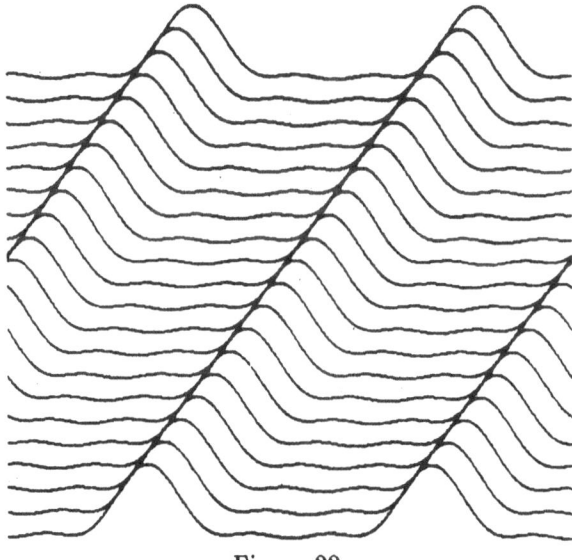

Figure 22:

Evolution of the wave profile with $N = 3$ from Fig.21 according to the dynamical equations (44). As expected, the wave travels undisturbed in shape, at constant velocity. The calculations are performed with timestep $h = .005$, and the profiles are shown at intermediate time intervals $20h$, so for a total time interval of length 2. The shift during this time agrees with the given value of the velocity λ.

pure translation as claimed.

Analytic proof.

Of course, despite the numerical evidence given above -which may be more or less convincing depending on ones attitude- the actual proof of the statement should follow from analytic arguments. These arguments are in fact quite involved when presented without any background information, but they are a straightforward application of general ideas about *relative equilibria* in Hamiltonian systems. For the sake of this presentation we will state each of the relevant steps. Each partial statement can be verified directly (with some effort), thereby producing in total a complete proof of the proposition.

Step 1.
Define the following function of the Fourier coefficients:

$$I(z) := \tfrac{1}{2} \sum_{k=1}^{k=N} (a_k^2 + b_k^2), \tag{49}$$

the sum of the squared amplitudes of the modes. We will call this function the *momentum*, for reasons to become clear. This function is a constant of the motion:

for solutions $z(t)$ of (32) it holds $\dfrac{d}{dt} I(z(t)) = 0$.

Step 2.
With the energy H and I as two constants of the motion for (32), consider the following *constrained optimisation problem*: for given γ positive, find the maximiser of H on the level set of I:

$$\text{Min } \{H(z)|I(z) = \gamma\}. \tag{50}$$

There exist solutions of this minimization problem, (one of) which we will call $\bar{z}$ (there exist also other critical points). This minimizer satisfies the equation found from applying Lagranges multiplier rule for constrained optimisation problems: for some "multiplier" $\lambda \in R$, $\bar{z}$ satisfies

$$\nabla H(\bar{z}) = \lambda \nabla I(\bar{z}). \tag{51}$$

This minimization problem can be solved numerically (if necessary, by hand for $N = 3$). For $N = 3$ one solution is given by the data (approximated to a few digits) in (48), where λ is the multiplier and the Fourier coefficients form the vector $\bar{z}$, when for the value of the constrained is taken (rather arbitrarily) $\gamma = 2\pi$.

Step 3.
When a periodic function $u(x)$ is shifted over some distance ϵ say, the Fourier coefficients of $u(x + \epsilon)$ are related in a simple way to those of $u(x)$. This defines a mapping, which we will call T_ϵ, and which is determined from:

if $z = (a_1, ..., b_n)$: $u(x) = \sum (a_k \cos kx + b_k \sin kx)$,

then $T_\epsilon z$ follows from :

$$u(x + \epsilon) = \sum a_k \cos k(x + \epsilon) + b_k \sin k(x + \epsilon) \tag{52}$$

$$= \sum (a_k \cos k\epsilon + b_k \sin k\epsilon) \cos kx + (-a_k \sin k\epsilon + b_k \cos k\epsilon) \sin kx. \tag{53}$$

It can be verified that T_ϵ can also be interpreted in the following way: $T_\epsilon(z_0)$ is the solution starting at $t = 0$ at $z = z_0$ after time $t = \epsilon$ of the following equation:

$$\dot{z}(t) = -D\nabla I(\bar{z}(t)) = -Dz, \tag{54}$$

i.e. the same equation as (32), but now with the function I introduced above as the Hamiltonian. Hence $z(\epsilon) = T_\epsilon(z_0)$. For this reason T_ϵ is called the "flow" of the equation (54). It changes the Fourier coefficients in such a way that the function merely experiences a shift. It needs hardly any calculation to convince oneself that both the original Hamiltonian H and the momentum I are invariant for such a shift:

$$\text{for all } \epsilon: \quad I(T_\epsilon z) = I(z), \quad \text{and} \quad H(T_\epsilon z) = H(z). \tag{55}$$

(Just differentiate the expressions with respect to ϵ, and use the (essential!!) fact that I is a constant of the motion for (32)).

Step 4.

Now combine the facts found in Steps 2 and 3. Since $\bar{z}$ is a constrained minimizer, and since the values of both H and I do not change when applying T_ϵ, for each ϵ also $\bar{z}(\epsilon) := T_\epsilon \bar{z}$ is a minimizer and satisfies the same equation as $\bar{z}$ itself:

$$\nabla H(\bar{z}(\epsilon)) = \lambda \nabla I(\bar{z}(\epsilon)), \quad \text{for all } \epsilon. \tag{56}$$

(This explains why there are many more solutions of the minimization problem; for the solution given in (48) the sine-coefficients were taken identically zero from the onset, invoking the symmetry of the profile to be expected). Consequently, writing $\epsilon = c_0\lambda t$, the function $t \rightarrow \bar{z}(t)$ satisfies the equation

$$\dot{\bar{z}}(t) = -c_0\lambda D\nabla I(\bar{z}(t)) = -c_0 D\nabla H(\bar{z}(t)). \tag{57}$$

Consequently, $t \rightarrow \bar{z}(t)$ satisfies the required equation, and is at the same time merely a translation. This completes the proof of the proposition!

Remark. All steps above are equally true for the partial differential equation itself, and they are in fact a consequence of that. Since (39) has constant coefficients (independent of x), there is translation symmetry. Related to this, the functional $\bar{I}(u) = \int \frac{1}{2}u^2$ is a constant of the motion. The Fourier truncation of $\bar{I}$ is precisely the function I given by (49). Then with $\bar{H}$ given by (41) and $\bar{I}$ one can consider the optimisation problem akin to (50). With functional analytic tools it can be shown that a minimizer to this (infinite dimensional !!) minimization problem exists. This function has to satisfy the equation that replaces (51), which turns out to be precisely equation (47). (The constant σ is another multiplier from the restriction to functions with mean value zero.) Since the solution of (47) is the travelling wave profile, and λ its speed, there is a complete analogy with the Fourier truncation.

Remark. The key point for the existence of exact travelling waves in the discretised equations is that there are *two* constants of the motion, just as for the original equation (39). One, the energy H, is a result of the Hamiltonian character of the equations, a property that can also be obtained (with some effort) for finite difference discretisations. However, the existence of the other one, I in this case, is directly related to the fact that the continuous symmetry in the partial differential equation (39) is preserved as a *continuous* symmetry in the discretised equations. Here this is owing to the fact that we used Fourier mode truncations. Finite difference methods will not preserve this continuous translation symmetry, and have no additional constant of the motion in general.

Concluding, travelling waves can be retained after discretisation if analytical knowledge is exploited carefully.

5   Epilogue

Let us review the foregoing examples and see what can be learned from them.

The first example -the logistic equation- showed the phenomenon of chaos. This was seen to appear for the simple Euler discretisation, and for large time steps as a consequence of the nonlinearity. However, for other discretisation schemes -all of which are in some way or another a modifaction of Euler's method- the same phenomenon can be expected to happen in general, possibly at larger time steps. For this simple example the time step had to be taken unrealistically large to observe the chaos. However, as is clear from the example on wave equations, when dealing with discretisations of partial differential equations, many (coupled, nonlinear) equations have to be solved, each of which has its own time scale. A time step which is small for some of them (as for the lowest modes in the Fourier truncation) may turn out to be large for others (the higher Fourier modes).

For the second example -the harmonic oscillator- we investigated energy conservation. Even when the energy was not conserved at the exact value, it was shown that an unbounded increase, or dissipation to a trivial state, can be avoided. The phase error made in the calculation of each mode cannot be overcome so easily when dealing with more modes of different frequency. Consequently, when the solution is obtained from a superposition of several calculated modes as in the example of the wave equation, the actual profile may be far off from the exact profile: in a pointwise norm, the error will be quite large. Although this may give some uneasy feelings, it seems an inevitable consequence of the numerical calculations, and calls for another measure to determine the quality of the calculation.

One such measure is the constancy of the constants of the motion, such as H and I above. Of course, this gives only a very rough indication since these are global quantities, taking contributions of all modes together. A more precise measure is the "spectrum" of the solutions, i.e. the amplitude of each mode (very often the quantity $a_k^2 + b_k^2$ is called the "energy" of the k-th mode, and one talks about the energy-spectrum. In the KdV-equation, this terminology is not consistent with the physical energy, which is approximated by H.) In general, this spectrum will vary in time because of (nonlinear) mode interactions. Then some time average can be taken. When compared with a more precise calculation (for instance the double number of modes), this gives some indication whether the individual modes, and their interactions, are calculated reliably or not. For the travelling waves considered in section 4, the spectrum is time independent and approximates the spectrum of the partial differential equation as good as possible (given the number of modes) as a consequence of the optimization property.

Software.

All calculations were performed on a regular PC (with mathematical coprocessor). There are various simple software packages for investigating ordinary differential equations. We mention PHASER and PHASEPLANE (see the references; the software comes along with the book). DYNPAO is another package, developed by R.A. Posthumus, University of Groningen, and is available upon request. Figures 1,5,6,7,8,9,10,15,16,19, and 20 were prepared with Phaseplane, and Figures 2,3,and 4 with Dynpao (all are direct copies of the screen).

For the wave equation, Figures 13,14,17,18,21 and 22 were obtained with PASCAL programmes, written (adapted) for these examples by van Beckum.

In section 4 some analytical calculations are needed to obtain the Hamiltonian (44) and, by taking its gradient, to find the equations (43) These calculations can be performed by hand or with an algebraic software package, such as DERIVE, MAPLE or MATHEMATICA. In order to avoid mistakes in the calculations on paper and in typing the commands, I used both methods. The results with the software packages can be used, by copying without retyping, in the Pascal programmes and in the final text of this report. These simple calculations give experience and confidence for dealing in the same way with more difficult problems.

A software package WAVEPACK is being developed at this moment as part of a cooperation between the mathematics institutes of the University of Twente and Institut Teknologi Bandung. WAVEPACK is aimed to be useful for both didactical and for research purposes. It visualises elementary concepts like superposition, dispersion, groupvelocity and Fourier mode interactions. At the same time it also allows to calculate the propagation of waves for KdV- and other equations. A particular facility enables the calculation of exact wave profiles of a soliton (or cnoidal wave) from an optimization problem. Also the distortion of waves over uneven bottoms can be calculated, as well as the effects of damping (viscosity) and selfexcitation. At this moment (April '92) a preliminary version is available. Interested readers can get a copy of the software, together with an extensive manual (including theory), from the author (at costprice, until further notice).

Acknowledgement.

Most of these ideas have been developed in discussions with F. van Beckum. Despite his great skill and experience in numerical calculations of much more difficult problems, he always showed great interest in my simple examples and questions. I am also gratefull for his direct help, in particular with the Pascal programmes.

References

- F. van Beckum & E. van Groesen: *Spatial discretizations of Hamiltonian Wave Equations*, University of Twente Memorandum, 1992, to be published.

- H.W. Broer, F. Dumortier, S.J. van Strien & F. Takens: *Structures in Dynamics, Finite dimensional deterministic studies*, Vol.2, North-Holland

Studies in Mathematical Physics, Amsterdam, 1991.

- R.L. Devaney: *An introduction to Chaotic Dynamical Systems*, Benjamin/Cummings, 1986.

- B. Ermentrout: *Phase Plane, The Dynamical Systems Tool*, Version 3.0, Brooks/Cole, California 1990.

- E. van Groesen, F. van Beckum, S. Redjeki & W. Djohan: WAVEPACK, *computer software for wave equations, including manual and theory*. Preliminary version, April 1992. Available upon request: E. van Groesen, Applied Mathematics, Univ. of Twente, P.O.Box 217, 7500 AE Enschede, The Netherlands.

- J. Guckenheimer & P. Holmes:*Nonlinear Oscillations, Dynamical Systems, and Bifurcation of Vector Fields*, Springer Verlag, Berlin, 1983.

- H. Koçak: *PHASER Differential and Difference Equations through Computer Experiments*, Springer Verlag 1986.

- H.G. Schuster: *Deterministic Chaos*, Physik Verlag, Weinheim, 1984.

- J.B. Whitham: *Linear and Nonlinear Waves*, Wiley, New York, 1974.

An introduction to the Finite Element Method

J.J.I.M. van Kan

Faculty of Technical Mathematics and Informatics
Delft University of Technology
P.O. Box 5031, 2600 GA Delft, The Netherlands

1   Introduction

The finite element method (FEM) has grown from a civil engineering tool into a general method for solving partial differential equations. In this area it beats its competitors: the finite difference method (FDM) and the finite volume method (FVM), in that it is better suited to deal with complex geometries and difficult boundary conditions. As opposed to that, it usually is more difficult to apply and the resulting sets of equations have a more complicated structure.

The objective of *any* numerical method for solving boundary value problems is to replace the original problem by a set of algebraic equations. If the originating differential equation is *linear* the resulting set of algebraic equations will be linear also. We shall illustrate this idea first with an elementary example: we show how these techniques create an algebraic set of equations on a one dimensional boundary value problem. Thereafter, we will show how the FEM works on a more complicated example: the 2-dimensional Poisson equation on a general domain.

2   An elementary example

Let us consider the following 1-dimensional boundary value problem on the interval $(0, 1)$:

Problem 1 *Find $u(x)$, $x \in (0, 1)$ such that*

$$-\frac{d^2u}{dx^2} = f, \quad u(0) = 0, \quad \frac{du}{dx}(1) = 0. \tag{1}$$

This boundary value problem may be interpreted as the temperature equilibrium in a rod, in which at one end the temperature is fixed (by keeping the rod in an ice-bucket for example) and on the other end the rod is isolated. Internal heat sources are represented by f. To tackle this problem numerically all three methods mentioned above start out in the same way: they divide the interval in a number of subintervals, say N, which we take for the moment of equal

A. van der Burgh and J. Simonis (eds.), Topics in Engineering Mathematics, 37–60.
© 1992 *Kluwer Academic Publishers.*

length h, so $h = 1/N$. See fig. 1. At the boundaries of the i-th interval, given by $((i-1)h, ih)$, we consider the value of the variable u. These internal boundaries are called the *nodal points* and the corresponding function values the *nodal values*. These nodal points are denoted by x_i, the nodal values by u_i, $i = 0 \ldots N$.

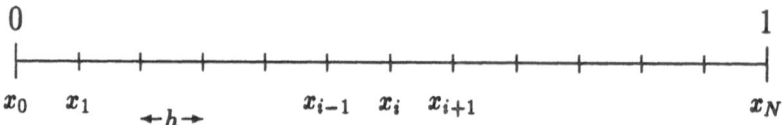

0   1

x_0 x_1 $\leftarrow h \rightarrow$ x_{i-1} x_i x_{i+1} x_N

Figure 1: Nodal points of interval (0,1)

2.1   Solution by Finite Difference Method

The finite difference method replaces the second derivative in a nodal point by a second divided difference:

$$\frac{d^2u}{dx^2}(x_i) \approx \frac{u_{i+1} - 2u_i + u_{i-1}}{h^2} \tag{2}$$

Substituting this into equation (1) for all internal nodal points gives the following set of equations:

$$-u_{i-1} + 2u_i - u_{i+1} = h^2 f_i, \quad i = 1, \ldots, N-1 \tag{3}$$

Since this is a set of $N-1$ equations and there are $N+1$ unknown u_i's, we are two equations short. These will be provided by the boundary conditions belonging to Problem 1. First of all the left hand boundary condition provides us with a value for u_0: $u_0 = 0$. So the first equation of the system (3) transforms into

$$2u_1 - u_2 = h^2 f_1 \tag{4}$$

The right hand side boundary condition can be approximated by a one-sided difference:

$$\frac{du}{dx}(1) \approx \frac{u_N - u_{N-1}}{h} = 0 \tag{5}$$

Adding this equation to system (3) we get

$$Au = h^2 f \tag{6}$$

in which A is an $N \times N$ tridiagonal matrix and $\boldsymbol{f}$ is an N-vector:

$$
A = \begin{pmatrix}
2 & -1 & 0 & \cdots & \cdots & 0 \\
-1 & 2 & -1 & \ddots & & \vdots \\
0 & \ddots & \ddots & \ddots & \ddots & \vdots \\
\vdots & \ddots & \ddots & \ddots & \ddots & 0 \\
\vdots & & \ddots & -1 & 2 & -1 \\
0 & \cdots & \cdots & 0 & -1 & 1
\end{pmatrix}, \quad
\boldsymbol{f} = \begin{pmatrix}
f_1 \\ f_2 \\ \vdots \\ f_{N-1} \\ 0
\end{pmatrix}
\tag{7}
$$

This set of equations has as many equations as unknowns and its solution can be thought of as an *approximation* of the solution of the original boundary value problem 1. The sources of error are of course the replacement of the second derivative by the second divided difference in the differential equation *and the replacement of the boundary condition by a one-sided divided difference.*

2.2   Solution by Finite Volume Method

The finite volume method integrates equation (1) over an interval $(x_i - \frac{1}{2}h, x_i + \frac{1}{2}h)$ to obtain:

$$
-\int_{x_i - \frac{1}{2}h}^{x_i + \frac{1}{2}h} \frac{d^2 u}{dx^2}\, dx = \int_{x_i - \frac{1}{2}h}^{x_i + \frac{1}{2}h} f\, dx, \quad i = 1, \ldots, N-1
\tag{8}
$$

The left hand side can be evaluated exactly, the right hand side by the midpoint rule:

$$
-\left(\frac{du}{dx}(x_i + \tfrac{1}{2}h) - \frac{du}{dx}(x_i - \tfrac{1}{2}h) \right) = h f_i
\tag{9}
$$

At the right side of the interval, we are left with an interval of size $\frac{1}{2}h$ which has to be dealt with separately:

$$
-\int_{x_N - \frac{1}{2}h}^{x_N} \frac{d^2 u}{dx^2}\, dx = \int_{x_N - \frac{1}{2}h}^{x_N} f\, dx
\tag{10}
$$

or

$$
-\left(\frac{du}{dx}(x_N) - \frac{du}{dx}(x_N - \tfrac{1}{2}h) \right) = \tfrac{1}{2}h f_N
\tag{11}
$$

If we replace $du/dx(x_i + \frac{1}{2}h)$ by central divided differences, in equation (9), that is

$$
\frac{du}{dx}(x_i + \tfrac{1}{2}h) = \frac{u_{i+1} - u_i}{h}
\tag{12}
$$

we obtain

$$
-\left(\frac{u_{i+1} - u_i}{h} - \frac{u_i - u_{i-1}}{h} \right) = h f_i, \quad i = 1, \ldots, N-1
\tag{13}
$$

in other words

$$- u_{i-1} + 2u_i - u_{i+1} = h^2 f_i, \quad i = 1, \ldots, N-1 \tag{14}$$

which is exactly the same as the set (3). However, the integral over the rightmost volume gives, since $u'_N = 0$

$$- u_{N-1} + u_N = \tfrac{1}{2} h^2 f_N \tag{15}$$

So again we end up with a set of equations of the form

$$A\boldsymbol{u} = h^2 \boldsymbol{f} \tag{16}$$

in which the matrix A is as in (7). The vector $\boldsymbol{f}$ however is a bit different:

$$\boldsymbol{f} = \begin{pmatrix} f_1 \\ f_2 \\ \vdots \\ f_{N-1} \\ \tfrac{1}{2} f_N \end{pmatrix} \tag{17}$$

This raises some doubts as to the correctness of either procedure, since they cannot both be right. We shall return to this problem in section 2.4.

2.3   Solution by Finite Element Method

The finite element method seeks to solve Problem 1 by an approximating function rather than by the nodal values. To this end, we introduce an approximating function $\tilde{u}$ defined by

$$\tilde{u}(x) = \sum_{i=0}^{N} u_i \phi_i(x) \tag{18}$$

in which u_i are again the nodal values and ϕ_i are interpolating functions, commonly known as *basis functions*. There is some freedom as to the choice of ϕ_i, which will in the end govern the accuracy of our approximation. For the moment we will not consider this issue but simply state, that we choose the ϕ_i such, *that on any subinterval $\tilde{u}$ is a linear function*. This makes $\tilde{u}$ look like a broken line as in fig. 2. What do the ϕ_i look like? Since $\tilde{u}$ is piecewise linear it is given in the i-th interval by

$$\tilde{u} = \frac{x_i - x}{h} u_{i-1} + \frac{x - x_{i-1}}{h} u_i, \quad x \in (x_{i-1}, x_i) \tag{19}$$

hence in the sum (18) a term containing u_i only contributes to the i-th and $i+1$-th subinterval. Usually these subintervals are called *elements*. We will denote the i-th element by e_i. We get for ϕ_i

$$\phi_i = \begin{cases} (x - x_{i-1})/h, & x \in e_i, \\ (x_{i+1} - x)/h, & x \in e_{i+1}, \\ 0 & \text{in all other elements} \end{cases} \tag{20}$$

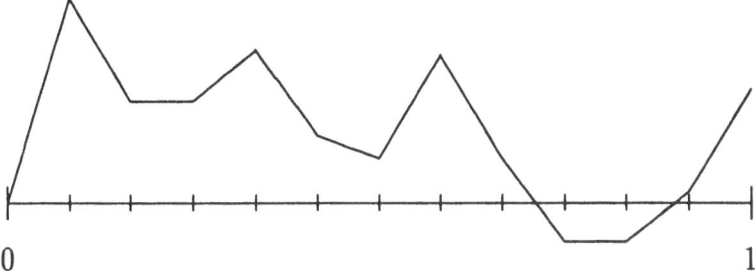

Figure 2: Example of typical $\tilde{u}$

A graphical representation of ϕ_i can be seen in fig. 3. For $i = N$ the basis

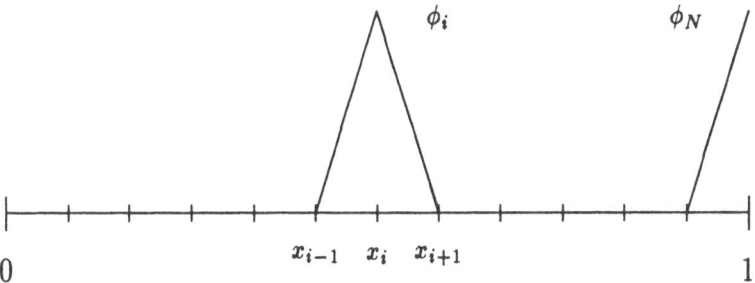

Figure 3: Graph of ϕ_i and ϕ_N

function ϕ_i is different:

$$\phi_N = \begin{cases} (x - x_{N-1})/h, & x \in e_N, \\ 0 & \text{in all other elements.} \end{cases} \qquad (21)$$

See fig. 3. In principle a similar basis function occurs at $x = 0$, but it does not enter the representation for $\tilde{u}$ by the imposed boundary condition $u(0) = 0$.

It will be clear, that with these definitions $\tilde{u}$ cannot satisfy equation (1) in any classical sense. The first derivative of $\tilde{u}$ is constant on an element and jumps across element boundaries, the second derivative is 0 on the interior of the elements and does not exist on element boundaries. We shall try to reformulate problem 1 in such a way that it makes sense for functions of type $\tilde{u}$.

2.3.1   Weak formulation of the problem

Let us take an arbitrary smooth function ψ, with $\psi(0) = 0$, multiply equation (1) left and right with ψ and integrate from 0 to 1. This yields :

$$-\int_0^1 \psi \frac{d^2 u}{dx^2}\, dx = \int_0^1 \psi f\, dx \qquad (22)$$

or, with partial integration

$$\int_0^1 \frac{d\psi}{dx} \frac{du}{dx}\, dx - \left[\psi \frac{du}{dx}\right]_0^1 = \int_0^1 \psi f\, dx \qquad (23)$$

Since $\psi(0) = 0$ and $du/dx(1) = 0$ this transforms into:

$$\int_0^1 \frac{d\psi}{dx} \frac{du}{dx}\, dx = \int_0^1 \psi f\, dx \qquad (24)$$

It will be clear, that any u that solves problem 1 will also satisfy equation (24), *whatever ψ we take*, as long as $\psi(0) = 0$. Let us now turn this around and consider the following problem:

Problem 2 *Find u, $u(0) = 0$, such that*

$$\int_0^1 \frac{d\psi}{dx} \frac{du}{dx}\, dx = \int_0^1 \psi f\, dx \qquad (25)$$

for any ψ with $\psi(0) = 0$.

It is implicitly understood that u and ψ are taken in such a way, that this formulation makes sense, that is, all integrals must exist. It can be shown (but we will not do this), that whenever both problems have a solution, these solutions coincide. We already saw, that a solution of problem 1 is also a solution of problem 2. But if problem 2 has a solution and problem 1 has not, this solution is called a *generalized* solution of problem 1. Problem 2 is called a *weak formulation* of problem 1, ψ is called a *testfunction* and the collection of all ψ's the *test space*. The collection of u for which problem 2 makes sense is called the *target space* of problem 2. We note, that the target space of problem 2 is essentially larger than the target space of problem 1: the first derivative must be (square) integrable in problem 2, whereas the second derivative must exist in problem 1. Moreover, $\tilde{u}$ of the previous section belongs to the target space of problem 2, so it is perfectly proper to look for solutions of problem 2 of the form $\tilde{u}$.

A question that may have risen in the mean time is: what happened to the boundary condition on the right side of the interval? The answer to that is, that it is incorporated into the weak formulation, that is, any u satisfying problem 2 automatically has $u'(1) = 0$. For that reason this type of boundary condition is called a *natural* boundary condition.

2.3.2   Choice of testfunctions. Galerkin's method

If we try to solve problem 2 by a function of type $\tilde{u}$ we encounter a problem: we can only vary N parameters in $\tilde{u}$, so we can satisfy equation (25) for only N testfunctions, whereas the testspace has of course an infinite number of testfunctions. But we have to remember, that we are looking for an *approximation* of the solution, so satisfying equation (25) for precisely N independent testfunctions should provide us with an approximation of the true solution. There is some freedom in the choice of the testfunctions, but there should be as many testfunctions as unknowns or *degrees of freedom* in the approximating function $\tilde{u}$. This provides us with a natural choice for the testfunctions: *choose every basisfunction that corresponds to an unknown degree of freedom as testfunction.* Hence we choose $\phi_i, i = 1, \ldots, N$ as testfunctions. This procedure is known as *Galerkin's method.* Substituting this and $\tilde{u}$ in equation (25) provides us with a finite element approximation:

Problem 3 *Find $u_1, u_2, \ldots, u_N$ such that*

$$\int_0^1 \frac{d\phi_k}{dx} \sum_{i=1}^N u_i \frac{d\phi_i}{dx} \, dx = \int_0^1 \phi_k f \, dx, \quad k = 1, \ldots, N \tag{26}$$

or interchanging the sum and the integral:

$$\sum_{i=1}^N u_i \int_0^1 \frac{d\phi_k}{dx} \frac{d\phi_i}{dx} \, dx = \int_0^1 \phi_k f dx, \quad k = 1, \ldots N. \tag{27}$$

This is a set of linear equations of the form:

$$S\boldsymbol{u} = \boldsymbol{f} \tag{28}$$

in which the matrix coefficients s_{ki} and the vector components f_k are given by

$$s_{ki} = \int_0^1 \frac{d\phi_k}{dx} \frac{d\phi_i}{dx} \, dx \tag{29}$$

$$f_k = \int_0^1 \phi_k f \, dx \tag{30}$$

For historical reasons S is called the *stiffness matrix*. It will be readily verified, that S is symmetric. We are left with the task of calculating the coefficients s_{ki} and the components f_k. We shall do this in a fashion that is typical for finite element approximations: by element matrices and element vectors.

2.3.3   Construction of S by element matrices

Consider the matrix coefficient s_{ki}

$$s_{ki} = \int_0^1 \frac{d\phi_k}{dx} \frac{d\phi_i}{dx} \, dx \tag{31}$$

$$= \sum_{j=1}^{N} \int_{x_{j-1}}^{x_j} \frac{d\phi_k}{dx} \frac{d\phi_i}{dx} \, dx \tag{32}$$

$$= \sum_{j=1}^{N} \int_{e_j} \frac{d\phi_k}{dx} \frac{d\phi_i}{dx} \, dx \tag{33}$$

So we have broken up the matrix coefficient s_{ki} into contributions from the various elements e_j. Another way of looking at this is, that the whole matrix may be built, by adding the contributions of every element e_j together. The beauty of this scheme is, that most contributions are 0. Consider the contribution of an element e_j to a matrix coefficient s_{ki}:

$$s_{ki}^{(j)} = \int_{e_j} \frac{d\phi_k}{dx} \frac{d\phi_i}{dx} \, dx \tag{34}$$

From figure (3), we see, that only ϕ_{j-1} and ϕ_j differ from 0 on e_j. Hence e_j *contributes only to* $s_{j-1,j-1}, s_{j-1,j}, s_{j,j-1}$ *and* $s_{j,j}$ *and all other contributions are 0.* These nonzero contributions are easily calculated:

$$s_{j-1,j-1}^{(j)} = \int_{e_j} \left(\frac{d\phi_{j-1}}{dx} \right)^2 dx \tag{35}$$

$$= \int_{x_{j-1}}^{x_j} \left(\frac{d}{dx} \frac{x_j - x}{h} \right)^2 dx \tag{36}$$

$$= \frac{1}{h} \tag{37}$$

and in the same way we find:

$$s_{j,j-1}^{(j)} = s_{j-1,j}^{(j)} = -\frac{1}{h}, \quad s_{j,j}^{(j)} = \frac{1}{h} \tag{38}$$

It is customary to write the nonzero contributions in matrix form:

$$S^{(j)} = \frac{1}{h} \begin{pmatrix} 1 & -1 \\ -1 & 1 \end{pmatrix} \tag{39}$$

and this is called the *element matrix* of the j-th element. In our example the element matrices of all elements are the same, with the exception of S^1 which consists of exactly one entry: $1/h$, which it contributes to $s_{1,1}$. This is not always the case, but it is quite usual, that the element matrices all have the same form and can be completely calculated from the values of the nodes belonging to the element. If that is the case, just *one* element matrix plus knowledge of which nodal points belong to which element *completely determines the matrix.*

Let us construct S. We start out with an $N \times N$ zero matrix and add the contribution of the first element:

$$
\begin{pmatrix}
1/h & 0 & \ldots & 0 \\
0 & 0 & \ldots & 0 \\
\vdots & & & \vdots \\
0 & \ldots & \ldots & 0
\end{pmatrix}
\tag{40}
$$

Next we add the contribution of e_2 which contributes to s_{11}, s_{12}, s_{21} and s_{22} giving:

$$
\begin{pmatrix}
2/h & -1/h & 0 & \ldots & 0 \\
-1/h & 1/h & 0 & \ldots & 0 \\
0 & 0 & \ldots & \ldots & 0 \\
\vdots & & & & \vdots \\
0 & \ldots & & \ldots & 0
\end{pmatrix}
\tag{41}
$$

Continuing in this way we finally obtain

$$
S = \frac{1}{h}
\begin{pmatrix}
2 & -1 & 0 & \ldots & \ldots & 0 \\
-1 & 2 & -1 & \ddots & & \vdots \\
0 & \ddots & \ddots & \ddots & \ddots & \vdots \\
\vdots & \ddots & \ddots & \ddots & \ddots & 0 \\
\vdots & & \ddots & -1 & 2 & -1 \\
0 & \ldots & \ldots & 0 & -1 & 1
\end{pmatrix}
= \frac{1}{h} A
\tag{42}
$$

in which A is just the matrix obtained by FDM and FVM methods. This is certainly encouraging.

2.3.4 Construction of f by element vectors

In the same way we may consider the components of the right hand side f_k:

$$
f_k = \int_0^1 \phi_k f \, dx
\tag{43}
$$

$$
= \sum_{j=1}^{N} \int_{e_j} \phi_k f \, dx
\tag{44}
$$

Again we have broken up the vector component f_k into contributions of the various elements e_j. Again we note, that element e_j only contributes to components f_{j-1} and f_j, since only ϕ_{j-1} and ϕ_j differ from 0 on e_j. Writing the contributions of e_j in vector form provides us with the *element vector* of the

j-th element:

$$f^{(j)} = \begin{pmatrix} \int_{e_j} \phi_{j-1} f \, dx \\ \int_{e_j} \phi_j f \, dx \end{pmatrix} \tag{45}$$

We still have to calculate the integrals in (45) in a convenient way. To this end we use the *Newton-Cotes* method. Consider an integral over e_j with arbitrary integrand g:

$$\int_{e_j} g \, dx \tag{46}$$

In order to find a numerical approximation we interpolate g in the same way as $\tilde{u}$, that is on e_j we replace g by

$$g \approx g_{j-1} \phi_{j-1} + g_j \phi_j \tag{47}$$

Substituting this in the integral we obtain:

$$\int_{e_j} g \, dx \approx g_{j-1} \int_{e_j} \phi_{j-1} \, dx + g_j \int_{e_j} \phi_j \, dx \tag{48}$$

$$= \frac{h}{2} [g_{j-1} + g_j] \tag{49}$$

In these expressions g_j is shorthand for $g(x_j)$. Substituting this in (45) and noticing that $\phi_{j-1}(x_j) = 0$ and $\phi_j(x_{j-1}) = 0$ we obtain

$$f^{(j)} = \frac{h}{2} \begin{pmatrix} f_{j-1} \\ f_j \end{pmatrix} \tag{50}$$

When we now build the right hand side from all these contributions we obtain:

$$f = h \begin{pmatrix} f_1 \\ f_2 \\ \vdots \\ f_{N-1} \\ \frac{1}{2} f_N \end{pmatrix} \tag{51}$$

This concludes the finite element approximation of problem 1. Close inspection will reveal, that it is identical to the finite volume approximation!

2.4 Conclusions

If we compare the three approximations, we see that the finite difference method differs from the other two in the last equation, having a right hand side of 0, where the other two have $\frac{1}{2} h^2 f_N$. So we may suspect that something is wrong with our finite difference approximation. What happened? As you may recall, we got this last equation from an approximation of the right side boundary condition:

$$\frac{du}{dx}(1) \approx \frac{u_N - u_{N-1}}{h} \tag{52}$$

and this is too crude. If we expand u in a Taylor series around $x = 1$, we find:

$$u(x_{N-1}) = u(1) - h\frac{du}{dx}(1) + \frac{h^2}{2!}\frac{d^2u}{dx^2}(1) + O(h^3) \tag{53}$$

and we see that approximation (52) corresponds with truncation of this series after the second term. If we had also retained the third term, we would have obtained, using equation (1):

$$u(x_{N-1}) = u(1) - h\frac{du}{dx}(1) - \frac{h^2}{2}f(1) \tag{54}$$

$$= u(1) - \frac{h^2}{2}f(1) \tag{55}$$

and corresponds with equation (15) as the FVM and FEM obtained. So the first conclusion is, that *natural* boundary conditions have to be treated with extreme care in a FDM, whereas they present no problem at all in a FVM or FEM.

Superficially it seems, that of the three methods FVM is superior, since it yields the approximation with the least amount of work. Although this is certainly true in this simple example, FDM and FVM are only effective on rectangular domains in higher dimensions, so their scope is severely limited. Also if the grid is not equidistant FDM and FVM run into all sorts of trouble, whereas the FEM extension to irregular grids is straightforward. The second conclusion therefore is, that FEM is superior to FVM and FDM on irregular geometries, especially in higher dimensional spaces.

Exercise

Let the grid spacing of $(0,1)$ be irregular, such that the length of element e_j is h_j. Show that the element matrix of e_j for problem 2 is given by:

$$S^{(j)} = \frac{1}{h_j}\begin{pmatrix} 1 & -1 \\ -1 & 1 \end{pmatrix} \tag{56}$$

and the element vector by

$$f^{(j)} = \frac{1}{2}h_j\begin{pmatrix} f_{j-1} \\ f_j \end{pmatrix} \tag{57}$$

$\square$

Finally we have seen, that in order to obtain a finite element approximation we need a weak formulation of the original problem. This is not entirely a trivial matter, since the test space has to satisfy certain conditions. A very lucid explanation on how to construct test spaces, (and a very good description of the FEM for that matter) can be found in [Hughes, 1987]. For the moment we just state, that for second order boundary problems testfunctions have to be zero on those parts of the boundary where there is no natural boundary condition.

3   The 2-dimensional Poisson equation

We shall now pay attention to a more complicated example which essentially is a straightforward two dimensional generalization of our problem 1. Let us consider a bounded region $\bar{\Omega} \subset I\!\!R^2$ with interior Ω and boundary Γ. This boundary is divided into two parts: Γ_1 and Γ_2. The outward unit normal (that is a vector of unit length, orthogonal to the boundary) is denoted by n. See fig. 4. On this region we consider the following boundary value problem:

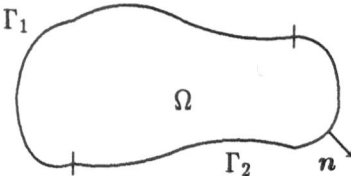

Figure 4: The region Ω.

Problem 4

$$- \text{div grad } u = f, \quad x \in \Omega, \tag{58}$$

$$u = 0, \quad x \in \Gamma_1, \tag{59}$$

$$\frac{\partial u}{\partial n} = 0, \quad x \in \Gamma_2 \tag{60}$$

in which

$$\frac{\partial u}{\partial n} = \frac{\partial u}{\partial x_1} n_1 + \frac{\partial u}{\partial x_2} n_2 \tag{61}$$

is the *directional derivative* in the direction of the outward normal, usually abbreviated to *normal derivative*.

Before we take the plunge, let us first consider the physics. First of all, we recall the divergence theorem, (also known as Gauss's integral theorem)

$$\int_V \text{div } q \; dV = \oint_S q.n \; dS \tag{62}$$

for any smooth vector field q and any volume V with boundary S. In this expression

$$\text{div}\, q \;=\; \frac{\partial q_1}{\partial x} + \frac{\partial q_2}{\partial y} \tag{63}$$

$$q.n \;=\; q_1 n_1 + q_2 n_2 \tag{64}$$

That means, that a vector field satisfying

$$\text{div}\, q = f, \quad x \in \Omega \tag{65}$$

has for all volumes $V \subset \Omega$

$$\int_V f \; dV = \oint_S q.n \; dS \tag{66}$$

The left hand side of this equality represents the *net* flux through the boundary of the volume. In other words, if $f = 0$ locally, the net flux vanishes and whatever goes in also goes out. *The flow is incompressible.* If $f \neq 0$ it means that there is a local *source*.

A flow of the form

$$q = -\text{grad}\, u = \left(\begin{array}{c} \partial u/\partial x \\ \partial u/\partial y \end{array} \right) \tag{67}$$

is typical for flows driven by a *potential*, like electrical current, heat flow (where the temperature acts as a potential) and ground water flow (where the hydrostatic pressure acts as a potential). The flow is directed from high potentials to low potentials, hence the minus sign.

So our problem 4 is a fairly general technical problem, representing an equilibrium in a potential driven flow. The right hand side f represents electrical charges, heat sources or water sources, as the case may be. The meaning of the boundary values is that the potential is kept constant at Γ_1 and there is no flow through Γ_2, that is the region is electrically or thermally insulated, or impenetrable at Γ_2. The boundary condition at Γ_2 is a *natural* boundary condition.

We may (and it is often done) rephrase the problem by substituting the expressions for div and grad:

$$-\left(\frac{\partial^2 u}{\partial x^2} + \frac{\partial^2 u}{\partial y^2} \right) = f \tag{68}$$

which is often abbreviated to

$$-\Delta u = f \tag{69}$$

and Δ is called the *Laplace* operator. We shall stick, however, to our formulation (58), since this is more transparent physically.

3.1   Weak formulation of the Poisson problem

To find a weak formulation we introduce again a test space of smooth functions ψ. Bearing in mind what has been said about the boundary values for ψ, we take $\psi = 0$ on Γ_1, since there is no natural boundary condition on Γ_1. We may formalize the definition of the test space $\mathcal{T}$ as follows:

$$\mathcal{T} = \left(\psi \; \middle| \; (\psi(\boldsymbol{x}) = 0, \quad \boldsymbol{x} \in \Gamma_1), \int_\Omega \|\text{grad }\psi\|^2 \; d\Omega < \infty \right) \tag{70}$$

the latter condition being a technicality, which need not concern us here. Before we can carry on we need a lemma, that comes straight from the divergence theorem, and that essentially provides a formula for partial integration in two dimensions:

Lemma 1 *For all $\psi \in \mathcal{T}$ and all smooth vector fields q*

$$\int_\Omega \psi \text{div } \boldsymbol{q} \; d\Omega = - \int_\Omega (\text{grad }\psi).\boldsymbol{q} \; d\Omega + \int_{\Gamma_2} \psi \boldsymbol{q}.\boldsymbol{n} \; d\Gamma \tag{71}$$

Proof.
From the divergence theorem (62) we have:

$$\int_\Omega \text{div}\,(\psi\boldsymbol{q}) \; d\Omega = \oint_\Gamma \psi\boldsymbol{q}.\boldsymbol{n} \; d\Gamma \tag{72}$$

Now recall that

$$\text{div}\,(\psi\boldsymbol{q}) = \psi \text{div } \boldsymbol{q} + (\text{grad }\psi).\boldsymbol{q} \tag{73}$$

and that $\psi = 0$ on Γ_1 and the lemma follows. $\square$

Now we are in a position to obtain a weak formulation of problem 4. We multiply equation (58) left and right by $\psi \in \mathcal{T}$, integrate over Ω and apply lemma 1 to obtain:

$$-\int_\Omega \psi \text{div grad } u \; d\Omega = \int_\Omega (\text{grad }\psi).(\text{grad } u) \; dV - \int_{\Gamma_2} \psi\boldsymbol{n}.\text{grad } u \; d\Gamma \tag{74}$$

$$= \int_\Omega \psi f \; d\Omega \tag{75}$$

Since

$$\frac{\partial u}{\partial n} = \boldsymbol{n}.\text{grad } u = 0, \quad \boldsymbol{x} \in \Gamma_2 \tag{76}$$

the boundary integral vanishes and we get:

$$\int_\Omega (\text{grad }\psi).(\text{grad } u) \; d\Omega = \int_\Omega \psi f \; d\Omega, \quad \forall \psi \in \mathcal{T} \tag{77}$$

which relation holds for the solution u of problem 4. We again turn this around in a now familiar fashion and give a weak formulation:

Problem 5 *Find $u \in T$ such that*

$$\int_\Omega (\text{grad } \psi).(\text{grad } u) \, d\Omega = \int_\Omega \psi f \, d\Omega, \quad \forall \psi \in T \tag{78}$$

We see that the target and test spaces coincide. This is accidental and has something to do with the fact that the boundary values on Γ_1 are *homogeneous*. Again the target space for problem 5 is essentially larger than that of problem 4 and the solution of problem 5 is a *generalized* solution of problem 4.

3.2   Galerkin's method for the Poisson problem

To apply Galerkin's method on problem 5 we take a set of *basis functions* $\phi_1, \phi_2, ..., \phi_N$ with the properties

1. $\phi_k \in T$, $k = 1, 2, \ldots, N$, in other words ϕ_k must belong to the target space.

2. (Approximation property) Every $v \in T$ can be approximated with acceptable accuracy by a linear combination of ϕ_k's.

Especially property 2 is formulated sufficiently hazy so as not to hamper us, but the idea should be clear: the better T can be approximated by the basis functions, the better our final approximation is going to be. For the moment we will not go further into the nature of the basis functions, but just let again

$$\tilde{u}(x) = \sum_{i=1}^{N} u_i \phi_i(x) \tag{79}$$

and take the basis functions as test functions to obtain a Galerkin approximation:

Problem 6 *Find $u_1, u_2, \ldots, u_N$ such that*

$$\int_\Omega (\text{grad } \phi_k).(\text{grad } \sum_{i=1}^{N} u_i \phi_i) \, d\Omega = \int_\Omega \phi_k f \, d\Omega, \quad k = 1, \ldots, N \tag{80}$$

or rearranging:

$$\sum_{i=1}^{N} u_i \int_\Omega (\text{grad } \phi_k).(\text{grad } \phi_i) \, d\Omega = \int_\Omega \phi_k f \, d\Omega, \quad k = 1, \ldots, N \tag{81}$$

So again we end up with a set of equations of the form

$$Su = f \tag{82}$$

with

$$s_{ki} = \int_\Omega (\operatorname{grad} \phi_k).(\operatorname{grad} \phi_i)\, d\Omega \tag{83}$$
$$f_k = \int_\Omega \phi_k f\, d\Omega \tag{84}$$

A good question at this point is, what has the finite element method to do with all this? The answer is: nothing yet. Galerkin's method is quite general for any choice of basis functions satisfying properties (1) and (2). It only becomes a finite element method by the special choice of basis functions like in section 2. The finite element method actually is a special case of Galerkin's method.

3.3   FEM approximation with linear triangular elements

The basis functions of a FEM approximation have a special nature. If we look at the elementary example of section (2), we see, that these basis functions are nonzero only on two elements, and are zero on the rest of the interval. This is typical of FEM basis functions. Generally it is true, that they are nonzero on a few elements only. The jargon for this phenomenon is, that they have *local support*. In our 2-dimensional problem we shall look at a simple example. We divide the region Ω into triangles, like in fig. 5. On a triangle e_j we approximate

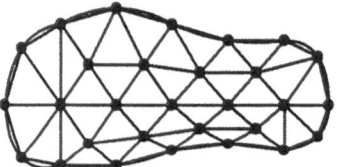

Figure 5: Triangulation of Ω.

u by a linear function. Suppose element e_j has vertices $x_0^{(j)}, x_1^{(j)}$ and $x_2^{(j)}$. On e_j we write

$$\tilde{u}(x) = u_0^{(j)}\lambda_0^{(j)}(x) + u_1^{(j)}\lambda_1^{(j)}(x) + u_2^{(j)}\lambda_2^{(j)}(x) \tag{85}$$

The functions $\lambda_k^{(j)}, k = 0, 1, 2$ are linear functions with the property

$$\lambda_k^{(j)}(x_k^{(j)}) = 1 \tag{86}$$
$$\lambda_k^{(j)}(x_i^{(j)}) = 0, \quad i \neq k \tag{87}$$

A graph of these functions can be found in fig. 6. The values $u_k^{(j)}, (k = 0, 1, 2)$

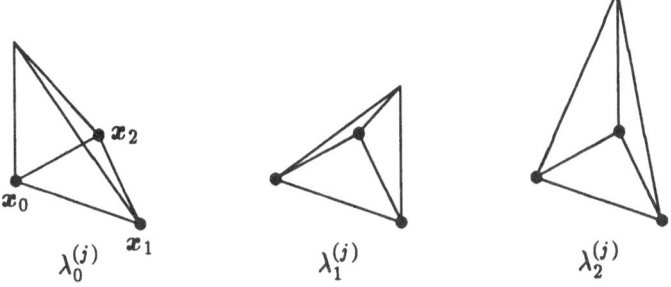

Figure 6: Graph of $\lambda_k^{(j)}$.

represent the *nodal values* in the vertices. These parameters will have to be determined in the end. Let us take a look at the nodal point x_i. The basis function ϕ_i for that nodal point is an assembly of various λ's on adjacent elements like in figure 7. Let us now proceed in the same fashion as in section 2.3.3 and 2.3.4.

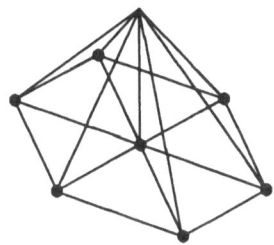

Figure 7: Basisfunction ϕ_i

To calculate the matrix entries of our Galerkin approximation we write

$$s_{ki} = \int_\Omega (\text{grad}\,\phi_k).(\text{grad}\,\phi_i)\, d\Omega \qquad (88)$$

$$= \sum_{j=1}^{N_e} \int_{e_j} (\text{grad}\,\phi_k).(\text{grad}\,\phi_i)\, de_j \qquad (89)$$

in which N_e stands for the total number of elements. Again we consider the contribution of an element e_j to the matrix S and we note the following. Let the nodal points of e_j have indices i_0, i_1 and i_2, then only ϕ_{i_0}, ϕ_{i_1} and ϕ_{i_2} are nonzero on e_j and they are equal to $\lambda_0^{(j)}, \lambda_1^{(j)}$ and $\lambda_2^{(j)}$ respectively. So the element e_j contributes to exactly *nine* matrix coefficients and this contribution is given by:

$$s_{i_n i_m}^{(j)} = \int_{e_j} (\text{grad}\,\lambda_n^{(j)}).(\text{grad}\,\lambda_m^{(j)})\, de_j, \quad n,m = 0,1,2. \qquad (90)$$

This all makes up for a lot of indexing, and this expression is usually simplified, by dropping the indices j and i. Expression (90) is then written as:

$$s^e_{nm} = \int_e (\operatorname{grad} \lambda_n).(\operatorname{grad} \lambda_m)\, de, \quad n, m = 0, 1, 2 \tag{91}$$

This is called a standard element formulation. It must be borne in mind, however, that this is really a shorthand notation for equation (90). In other words, all vertices and $\lambda's$ that enter in this expression must be taken with respect to the actual element we are considering. In the same way we find contributions to the right hand side:

$$f^e_n = \int_e \lambda_n f\, de, \quad n = 0, 1, 2. \tag{92}$$

To find the matrix S and the right handside f these contributions have to be added in the proper places. To do that we only need to know which nodes belong to which elements. The algorithm becomes quite simple then: for all elements calculate the contributions and add them up. We will not do this. That is what computers are for.

3.4   Calculation of λ

To finish the problem we have to know the expression for $\lambda_n, n = 0, 1, 2$. Consider the standard element e with vertices x_0, x_1, x_2. We try to find λ_n of the form:

$$\lambda_n(x) = a^n_0 + a^n_1 x + a^n_2 y, \quad n = 0, 1, 2 \tag{93}$$

with the property

$$\lambda_n(x_n) \;=\; 1 \tag{94}$$
$$\lambda_n(x_m) \;=\; 0, \quad m \neq n \tag{95}$$

That is, we have to solve the matrix equation:

$$\begin{pmatrix} 1 & x_0 & y_0 \\ 1 & x_1 & y_1 \\ 1 & x_2 & y_2 \end{pmatrix} \begin{pmatrix} a^0_0 & a^1_0 & a^2_0 \\ a^0_1 & a^1_1 & a^2_1 \\ a^0_2 & a^1_2 & a^2_2 \end{pmatrix} = \begin{pmatrix} 1 & 0 & 0 \\ 0 & 1 & 0 \\ 0 & 0 & 1 \end{pmatrix} \tag{96}$$

or in matrix-vectorform:

$$XA = I \tag{97}$$

in other words,

$$A = X^{-1} \tag{98}$$

and we just have to invert the matrix X. This is straightforward and, letting $\det(X) = \Delta$ we obtain:

$$a_0^0 = \frac{(x_1 y_2 - x_2 y_1)}{\Delta}, \quad a_1^0 = \frac{y_1 - y_2}{\Delta}, \quad a_2^0 = \frac{x_2 - x_1}{\Delta} \tag{99}$$

$$a_0^1 = \frac{(x_2 y_0 - x_0 y_2)}{\Delta}, \quad a_1^1 = \frac{y_2 - y_0}{\Delta}, \quad a_2^1 = \frac{x_0 - x_2}{\Delta} \tag{100}$$

$$a_0^2 = \frac{(x_0 y_1 - x_1 y_0)}{\Delta}, \quad a_1^2 = \frac{y_0 - y_1}{\Delta}, \quad a_2^2 = \frac{x_1 - x_0}{\Delta} \tag{101}$$

Note that coefficients in the same column all have the same structure and can be obtained from one another by cyclic permutation. Putting all this into the expression of s_{nm}^e gives

$$s_{nm}^e = \int_e (\operatorname{grad} \lambda_n).(\operatorname{grad} \lambda_m) \, de \tag{102}$$

$$= \int_e \left(\frac{\partial \lambda_n}{\partial x} \frac{\partial \lambda_m}{\partial x} + \frac{\partial \lambda_n}{\partial y} \frac{\partial \lambda_m}{\partial y} \right) de \tag{103}$$

$$= \int_e (a_1^n a_1^m + a_2^n a_2^m) \, de \tag{104}$$

$$= \tfrac{1}{2}|\Delta|(a_1^n a_1^m + a_2^n a_2^m) \tag{105}$$

the latter equality coming from the fact, that $|\Delta|$ is just twice the area of the triangle. Note that the coefficients only involve differences of vertex coordinates and hence are invariant under translation.

Without proof we give the Newton Cotes approximation of an arbitrary integral over an element:

$$\int_e g \, de \approx \tfrac{1}{6}|\Delta|(g_0 + g_1 + g_2) \tag{106}$$

which enables us to find the element vector:

$$f_n^e = \int_e \lambda_n f \, de = \tfrac{1}{6}|\Delta|f(x_n), \quad n = 0, 1, 2 \tag{107}$$

since $\lambda_n(x_m) = 0, n \neq m$.

3.5   Solution of the equations

After formulating the element matrices and element vectors, a lot of work remains to be done, that falls a bit outside the scope of this presentation. First of all the matrix S and the vector f have to be assembled. This is essentially a software problem and there are various finite element packages that provide structures to do just that.

Secondly, the remaining system has to be solved. This may be quite a task, especially when the problem has many variables, as in 3-dimensional calculations. The matrix we obtained in the one-dimensional problem was *tridiagonal*. In our two-dimensional problem we get a similar structure of the matrix characterized by the following properties:

- On every row most matrix entries are zero.

- The non zero entries are clustered around the diagonal

This kind of matrices are called *sparse*. Various methods to solve this type of problems actually use the sparsity of the matrix to reduce the number of operations. Roughly one may state, that the better the nonzero matrix entries are clustered towards the diagonal, the less the number of operations one is going to need to solve the system by direct methods as Gaussian elimination or the like. This clustering is governed by the indices of the respective unknowns. This can make quite a difference as even our simple example of fig. 5 shows. We have only 20 unknowns in this case, but the optimal numbering has no nonzero entry more than four places away from the diagonal (one says that the *profile width* is at most four), whereas a less optimal numbering can get a profile width of ten. See fig. 8. But a number of 20 unknowns is fairly atypical. In ordinary

Figure 8: Matrix profile for optimal and non-optimal indexing. • denote nonzero entries.

life hundreds of unknowns is no exception and a lot depends on whether the indexing of these unknowns is close to optimal. To obtain the *best possible* indexing is a combinatorial problem the solution of which is much more difficult than the problem we started out to solve. Fortunately we do not need the best possible indexing, and there are ways [Cuthill-McKee] to obtain an indexing that is close to optimal.

Sparse matrices are also well suited to *iterative* methods, like CGS [Meijerink-Van der Vorst]. In that case the indexing is not critical.

Finally an error estimate of the approximating solution should be made, because otherwise it is of no use. Under fairly general conditions there is a result that states, that the accuracy of the solution is of the same order as the best possible approximation of the exact solution by the basis functions. This is a very loose description of a result that is actually fairly deep, but for the moment it has to do. We refer to [Hughes], [Strang-Fix] for further reading on this very interesting subject.

3.6 Beyond linear elements

With the last statement of the preceding section in mind a natural question would be, whether we cannot do better than linear approximation. The answer is yes, but at a price. For instance, if we take an approximation with piece-wise quadratic polynomials, we need 6 coefficients per element. (A bivariate quadratic polynomial has 6 terms.) Moreover, we must take care that these piecewise quadratic polynomials are continuous across element boundaries, be-cause otherwise integral (83) makes no sense. Fortunately this is easy to achieve. By positioning three unknown function values in the vertices of an element and three unknowns in the middle of the edges, we see that the function at an edge is determined by three function values. And since three parameters completely determine a quadratic in one dimension two adjacent elements have the same function value at their common edge.

There are some more problems at the boundary. In our linear approxima-tion we just approximated the boundary by straight lines. We cannot do this in our quadratic approximation, because otherwise we would lose an order of accuracy and that would make the whole exercise pointless. Hence we must also approximate the boundary by piecewise quadratic polynomials. In fact, this causes the 'triangles' at the boundary to have one curved edge and this makes the evaluation of integral (83) very complicated.

We also have to take care, that the numerical integration rule with which we calculate the element vector is sufficiently accurate.

So it can be done, but it is not easy. Even higher order of approximation is possible, with comparable higher order of complexity. This, however, is far beyond the scope of this introduction.

4   Final remarks

We have seen the finite element method at work on two examples that were sufficiently simple so as not to drown in technical detail. Yet they give a flavour of the main characteristics of the method, which are

- It works on general domains.

- A weak formulation is needed.

- This is approximated by a finite number of basis/test functions *of local support*.

- The contributions calculated for a standard element completely determine the set of equations.

- A supporting software structure is needed to assemble the matrix and vector.

- A supporting software structure is needed to solve the resulting set of equations.

Commercially available FEM packages usually restrict themselves to problems from a specific discipline (like elasticity, hydrodynamics etc.) and they require almost no work or even understanding from the user. But their scope is limited and their structure usually rigid. On the other hand, there are packages floating around in the scientific community that are almost completely general. They, however, require a deeper understanding of the user, and must be handled with care.

Further reading.

Over the years an impressive amount of literature has appeared about the finite element method. Of the many titles we can only name a few, but this list covers most of the important aspects:

- engineering

- programming

- mathematics

First of all we mention the series of *Becker et al.*, which consists of 7 volumes and covers all three aspects thoroughly. The series starts out with an elementary introduction (about the level of this paper) and gradually deepens and broadens the view, covering also computational and mathematical aspects. The last two volumes are dedicated to solid mechanics and fluid mechanics respectively.

Covering specific engineering aspects are the books by *Cuvelier et al.* (fluid mechanics), *Kikuchi* (solid mechanics) and *Zienkewicz* (most structural analysis, but something about fluids). These books are fairly well suited for engineers who do not want to be bothered too much by the mathematical ins and outs of the FEM. Kikuchi and Zienkewicz are straightforward 'how to' books, Cuvelier has also a chapter on the mathematical theory.

There are many problems involved in actually programming a FEM. *Hinton and Owen* wrote a book devoted solely to that. There are many example programs (in FORTRAN) and the book is an absolute must for anyone aspiring to write a finite element package. Whether this is a wise thing to do is another matter, considering the many packages that are available right now.

We already mentioned the book by *Hughes* that is a fairly complete introduction covering all three aspects.

About the mathematical aspects we mention three books. In increasing degree of mathematical involvement we have *Mitchell & Wait, Strang & Fix* and *Ciarlet*. For all three a basic knowledge of functional analysis is needed. Some knowledge of (Lebesgue) integration theory would not hurt either.

This list is by no means complete but in my opinion it covers most of what might be needed by someone who wants to dig deeper.

References

Becker, E.B., G.F. Carey, J. Tinsley Oden,
Finite Elements.
Vols I-VII
Prentice-Hall, Inc., Englewood Cliffs, New Jersey, 1981-1984.
Ciarlet, P.G.
Finite element methods for elliptic problems.
North Holland Publishing Co, Amsterdam, 1986.
Cuthill, E., J. McKee,
Reducing the bandwidth of sparse symmetric matrices.
Proc. ACM Nat. Conf., New York 1969, 157-172.
Cuvelier, C., A.Segal, A.A van Steenhoven,
Finite element method and Navier Stokes equations.
D. Reidel Publishing Company, Dordrecht, 1986.
Hinton, E., D.R.J. Owen,
Finite element programming.
Academic Press, New York, 1977.
Hughes, Thomas J.R.,
The Finite Element Method.
Prentice-Hall, Inc., Englewood Cliffs, New Jersey, 1987.
Kikuchi, N.,
Finite element methods in mechanics.
Cambridge University Press, Cambridge, 1986.
Meijerink, J.A., H.A. van der Vorst,
An iterative solution method for linear systems of which the coefficient matrix is a symmetric M-matrix.
Math. Comp. **31**, 148-162, 1977.
Mitchell, A.R., R. Wait,
The finite element method in partial differential equations.
John Wiley, New York, 1977.
Strang, G., G.J. Fix,
An analysis of the finite element method.

Prentice-Hall, Inc., Englewood Cliffs, New Jersey, 1973.

Zienkiewicz, O.C.

The finite element method in engineering science.

McGraw-Hill, New York, 1971.

Coupling of
sound and structural vibrations

C. Kauffmann

Faculty of Technical Mathematics and Informatics

Delft University of Technology

P.O. Box 5031, 2600 GA Delft, The Netherlands

Abstract

Vibrational properties of hull plates of ships are considerably modified by the presence of the water. A simple model problem is presented that accounts for the fluid-loading, i.e. the coupling of the plate's flexural vibrations and the acoustic field in the fluid. This model, consisting of a thin, transversely vibrating plate with a compressible fluid at one side and excited by a time-harmonic load, is analyzed in some detail. The model equations are solved by means of a Green integral representation along with a set of coupled boundary integral equations. The kernel of the integral representation is found by using the Fourier integral transform technique and by evaluating numerically the inverse transform in the complex wavenumber plane. For the two-dimensional case the boundary integral equations degenerate to a system of algebraic equations, thus yielding an exact representation for the solution. Numerical results show the fluid-loading effect on the resonance frequencies of the plate, which are shifted downwards relative to the *in vacuo* natural frequencies, while acoustic radiation contributes to the damping of the plate's resonant modes.

1 Introduction

Dynamic interaction of fluids and structures is encountered in many practical and industrial situations. Both experimental and theoretical research has been done (and is still going on) in the following area's:

- vibration of fluid-filled pipes (like cooling water pipes of diesel engines and industrial plant piping systems)

A. van der Burgh and J. Simonis (eds.), Topics in Engineering Mathematics, 61–91.
© 1992 *Kluwer Academic Publishers.*

- liquid sloshing in flexible containers

- hydro-elastic vibrations of offshore structures

- vibro-acoustic behaviour of fluid-filled cavities (partially) bounded by a flexible wall (domed sonar systems, kettle drums, ...)

- vibration, radiation and scattering of marine structures

- flow-induced vibrations (flutter, aeroelasticity, ...)

In this paper a very simple model problem is discussed in order to deal with the vibro-acoustic behaviour of a typical hull plate of a ship. A qualitative description of the fluid-structure interaction mechanism can be given as follows. Consider a vibrating structure (consisting of plates, shells, ...) that is in continuous contact with a surrounding fluid. A pressure field present in the fluid will act upon the structure's surface as a driving force and thus generate structural vibrations. Conversely, continuous contact at the fluid-structure interface implies that the normal velocity component of the wet surface of the structure is imposed on the fluid. Moreover, due to the extended reaction of the structure

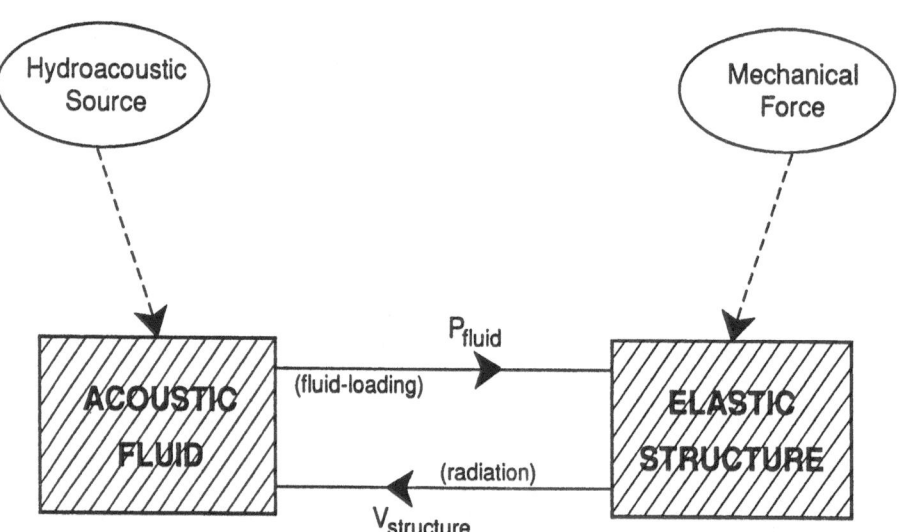

FIGURE 1. Fluid-structure interaction

to local forcing as well as the non-local reaction of the fluid to a local velocity source the coupling is of a non-local type. The primary source of the dynamic motion of this fluid-structure system may be of the mechanical kind, i.e. which acts directly on the structure or of the hydrodynamical kind, the latter effectively generates pressure and velocity disturbances in the fluid. In Figure 1 the interaction process is summarized schematically. Because of the doubly connected coupling mechanism one is confronted with the problem to solve the equations of motion for the structure and for the fluid simultaneously. The intuitive approach of separately solving the structural and the fluid problem and to link these solutions as a final step obviously ignores the interaction mechanism and therefor should be rejected.

A very successful numerical approach in solving this type of coupled problems was developed during the last decade. In this method the structure is modelled by finite elements, while the Kirchhoff-Helmholtz integral equation is used to introduce fluid boundary elements at the structure's surface. The coupling conditions provide the closure of the problem and a linear system of the following type is obtained.

$$
\begin{pmatrix} -M\omega^2 + K & -P \\ \rho_0\omega^2 C & H \end{pmatrix} \cdot \begin{pmatrix} u \\ p \end{pmatrix} = \begin{pmatrix} f_{mech} \\ q_{acoust} \end{pmatrix},
$$

where $-M\omega^2 + K$ describes the structural system, P is (almost) diagonal and takes into account the fluid-loading, while C and H are matrices obtained from the boundary element discretization of the Kirchhoff-Helmholtz integral equation for the pressure at the structure's wet surface. Alternatively, by employing variational techniques it is possible to obtain a symmetric formulation for the coupled finite element / boundary element equations. This method is very effective, especially because the use of boundary elements reduces the fluid domain dramatically. Moreover, it is not restricted to special (separable) geometries. However, the method is inherently restricted to the low-frequency range, while inaccuracies in the radiated field are likely to occur. Moreover, physical insight in the coupling mechanism is poor and there is still a need for analytical solutions, which can provide benchmarks. Last (but not least) analytical solutions may be used for extracting general design rules from a parameter study.

2 Governing equations and analysis

Description of the model

In this paper the structure is modelled by a thin, transversely vibrating, elastic plate. A perfect acoustic fluid occupies the halfspace above the plate. The plate is effectively infinite and it is backed by a vacuum. Fluid-loading is established through the acoustic pressure at the plate's surface, which drives the plate. On the other hand the out-of-plane velocity of the plate radiates sound into the fluid.

In addition, the plate is constrained along a contour, separating a finite part of the plate from its coplanar complement. The finite part of the

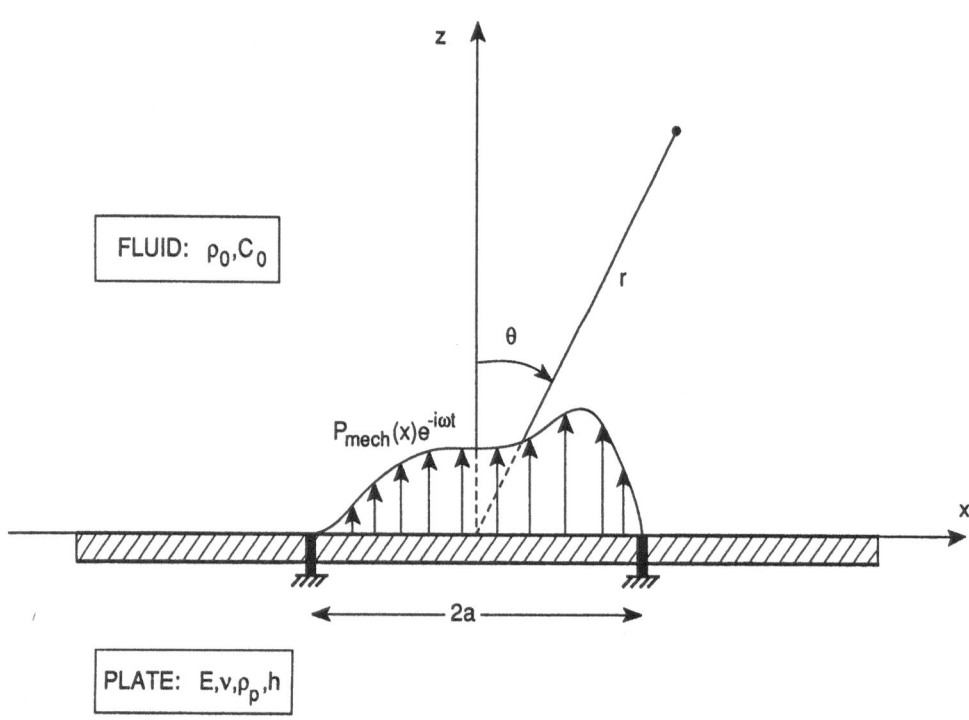

FIGURE 2. Reinforced, fluid-loaded plate driven by a time-harmonic load

plate has, of course, its *in vacuo* resonant behaviour. However, fluid-loading is expected to modify the resonance properties of the plate.

A Cartesian reference frame is defined in such a way that $z > 0$ corresponds to the fluid halfspace, while the plate occupies the layer $-h \leq z \leq 0$, where h denotes the plate's thickness. In order to simplify the analysis, this paper adresses the two-dimensional case, i.e. the acoustic field is supposed to be uniform with respect to one of the lateral co-ordinates, say y. Analogously, the plate response is uniform in the y-direction.

Basic equations

The out-of-plane flexural motion of the plate is governed by

$$D\frac{\partial^4 u}{\partial x^4} + \rho_p h \frac{\partial^2 u}{\partial t^2} = p_{mech}(x,t) - p(x,0,t), \tag{1}$$

where $u(x,t)$ is the deflection of the plate, p_{mech} is the mechanical force distribution per unit length (in y-direction) applied to the plate and $p(x,z,t)$ is the acoustic pressure in the fluid. $D = Eh^3/12(1-\nu^2)$ is the bending stiffness, where E is Young's modulus, ν is Poisson's ratio and ρ_p is the density of mass of the plate. The spatial co-ordinate is x and t is time.

The plate is constrained along two parallel line stiffeners located at $x = \pm a$. This means that the resonant behaviour of a strip of finite width and infinite length is introduced into the fluid-plate system. Two coplanar, semi-infinite plates of the same material and thickness are joined to the strip at $x = \pm a$. Boundary conditions at these line stiffeners must be specified. In this paper the clamped case is treated, i.e.

$$u(x,t) = 0 \text{ and } \frac{\partial u}{\partial x}(x,t) = 0, \text{ for } x \uparrow a, \ x \downarrow a, \ x \downarrow -a, \ x \uparrow -a. \tag{2}$$

Additionally, initial conditions are needed for $u(x,t_0)$ and $(\partial/\partial t)u(x,t_0)$, while causality requirements (as well as boundedness conditions) should be imposed in order to determine a unique solution.

The acoustic field in the fluid is governed by the linearized equations of motion for a perfect fluid, i.e.

$$\frac{\partial \rho}{\partial t} + \rho_0 \left[\frac{\partial v_1}{\partial x} + \frac{\partial v_3}{\partial z} \right] = 0, \tag{3a}$$

$$\frac{\partial v_1}{\partial t} = -\frac{1}{\rho_0} \frac{\partial p}{\partial x}, \quad \frac{\partial v_3}{\partial t} = -\frac{1}{\rho_0} \frac{\partial p}{\partial z}, \tag{3b, c}$$

$$p = c_0^2 \rho, \tag{3d}$$

where $\rho(x, z, t)$ and $v_{1,3}(x, z, t)$ denote the acoustic fluctuations of the density of mass and the velocity components respectively, while ρ_0 is the static density of mass and c_0 is the speed of sound. By eliminating ρ, v_1 and v_3 a wave equation can be found for p:

$$\frac{1}{c_0^2} \frac{\partial^2 p}{\partial t^2} - \left[\frac{\partial^2 p}{\partial x^2} + \frac{\partial^2 p}{\partial z^2} \right] = 0, \tag{4}$$

Similar equations hold for ρ and for v_1 and v_3.

If, eventually, hydroacoustic sources are present in the fluid, equation (4) becomes inhomogeneous, the right-hand-side representing the source. At present, the vibrating plate radiates sound into the fluid. This process is controlled by the boundary condition

$$-\frac{1}{\rho_0} \frac{\partial p}{\partial z}(x, 0, t) = \frac{\partial v_3}{\partial t}(x, 0, t) = \frac{\partial^2 u}{\partial t^2}(x, t), \quad z = 0, \tag{5}$$

which follows from (3c) and the continuity of the velocity at the plate's surface.

Again, initial conditions must be specified for $p(x, z, t_0)$ and for either $v_{1,3}(x, z, t_0)$ or $(\partial/\partial t)p(x, z, t_0)$. In connexion with the boundary condition (5), this will ensures a unique, causal solution.

Frequency domain analysis

Initial conditions however, will not be specified here, because practical interest is in the spectral properties of the fluid-plate system rather than in transient behaviour. So, all quantities of interest (driving force,

plate displacement, acoustic pressure) are supposed to vary harmoni-
cally with time, ω being the circular frequency. See Appendix A.
Introduction of appropriate scalings for both dependent and independent variables yields the following nondimensional forms for the plate
equation and boundary conditions:

$$\frac{d^4\bar{u}}{d\bar{x}^4} - \alpha^4\bar{u}(\bar{x}) = \bar{p}_{mech}(\bar{x}) - \bar{p}(\bar{x}, \bar{z} = 0), \quad -\infty < \bar{x} < \infty, \ \bar{x} \neq \pm 1, \quad (6)$$

$$\bar{u}(\bar{x}) = 0 \text{ and } \frac{d\bar{u}}{d\bar{x}}(\bar{x}) = 0, \text{ for } \bar{x} \uparrow 1, \ \bar{x} \downarrow 1, \ \bar{x} \downarrow -1, \ \bar{x} \uparrow -1. \quad (7)$$

At infinity, i.e. for $\bar{x} \to \pm\infty$, a radiation condition and a boundedness
requirement hold.
All nondimensional quantities are indicated by a bar (which will be
dropped) according to the following scalings:

$$\bar{x} = x/a, \ \bar{z} = z/a, \ \bar{u} = u/(F'a^3/D), \ \bar{p}_{mech} = p_{mech}/(F'/a), \ \bar{p} = p/(F'/a),$$

where F' is given by $F' = \int_{-a}^{a} p_{mech}(x)dx$. The acoustic pressure satisfies
the Helmholtz equation with a Neumann condition at $\bar{z} = 0$:

$$\bar{\Delta}\bar{p} + \beta^2\bar{p}(\bar{x}, \bar{z}) = 0, \quad -\infty < \bar{x} < \infty, \ \bar{z} > 0, \quad (8)$$

$$\frac{\partial\bar{p}}{\partial\bar{z}}(\bar{x}, 0) = \sigma^4\bar{u}(\bar{x}), \quad -\infty < \bar{x} < \infty, \ \bar{z} = 0, \quad (9)$$

where $\bar{\Delta}$ is the Laplace operator in nondimensional form, i.e. $\bar{\Delta} \equiv \partial^2/\partial\bar{x}^2 + \partial^2/\partial\bar{z}^2$. A radiation condition at infinity completes the set
of governing equations.
The nondimensional parameters α, β and σ are defined by

$$\alpha^4 = \frac{m\omega^2 a^4}{D}, \quad \beta = k_0 a = \omega a/c_0, \quad \sigma^4 = \frac{\rho_0\omega^2 a^5}{D}. \quad (10)$$

α and β are the nondimensional wavenumbers in the plate and the fluid
respectively, while σ accounts for the coupling. $m = \rho_p h$ is the mass
per unit area of the plate.

Solution by integral representations

The solution of (8), (9) is readily obtained formally, i.e. without specifying $\bar{u}(\bar{x})$, by using the Neumann Green function for the halfspace $\bar{z} > 0$: (see Appendix B)

$$\bar{p}(\bar{x}, \bar{z}) = \frac{1}{2i}\sigma^4 \int_{-\infty}^{\infty} H_0^{(1)}(\beta\sqrt{(\bar{x} - \bar{\xi})^2 + \bar{z}^2})\bar{u}(\bar{\xi})d\bar{\xi}. \qquad (11)$$

By substituting the integral representation (11) into the right-hand-side of (6) a fourth order integro-differential equation for $\bar{u}(\bar{x})$ is obtained:

$$\frac{d^4\bar{u}}{d\bar{x}^4} - \alpha^4\bar{u}(\bar{x}) = \bar{p}_{mech}(\bar{x}) - \frac{1}{2i}\sigma^4 \int_{-\infty}^{\infty} H_0^{(1)}(\beta|\bar{x}-\bar{\xi}|)\bar{u}(\bar{\xi})d\bar{\xi}, \quad \begin{array}{c} -\infty < \bar{x} < \infty, \\ \bar{x} \neq \pm 1. \end{array}$$
$$(12)$$

The boundary conditions (7) at $x = \pm a$ and the radiation condition and boundedness requirement at infinity determine a unique solution. Equation (11) is solved by applying Green's theorem to $\bar{u}(d^4/d\bar{x}^4)\bar{\gamma} - \bar{\gamma}(d^4/d\bar{x}^4)\bar{u}$, where $\bar{\gamma}$ is the fundamental solution for the fluid-loaded, infinite plate subject to line drive, i.e. the solution of

$$\frac{d^4\bar{\gamma}}{d\bar{x}^4} - \alpha^4\bar{\gamma}(\bar{x}; \bar{x}_0) = \delta(\bar{x} - \bar{x}_0) - \frac{1}{2i}\sigma^4 \int_{-\infty}^{\infty} H_0^{(1)}(\beta|\bar{x} - \bar{\xi}|)\bar{\gamma}(\bar{\xi}; \bar{x}_0)d\bar{\xi}, \quad (13)$$

on $-\infty < \bar{x} < \infty$, satisfying radiation and boundedness conditions at infinity. A discussion on $\bar{\gamma}(\bar{x}; \bar{x}_0) \equiv \bar{\gamma}(\bar{x} - \bar{x}_0)$ is given in Appendix C. It is easy to show that terms that represent the fluid-loading in (12) and (13) cancel if Green's theorem is applied on the domain $-\infty < \bar{x} < \infty$. Special attention must be paid to the points $\bar{x} = \pm 1$, where the second and third order derivative of $\bar{u}$ are discontinuous, while $\bar{u}$ and its first order derivative are continuous. The result is

$$\bar{u}(\bar{x}) = \bar{u}_\infty(\bar{x}) + \bar{\lambda}_1(1)\bar{\gamma}(1 - \bar{x}) - \bar{\lambda}_1(-1)\bar{\gamma}(-1 - \bar{x})$$
$$-\bar{\lambda}_2(1)\bar{\gamma}'(1 - \bar{x}) + \bar{\lambda}_2(-1)\bar{\gamma}'(-1 - \bar{x}), \qquad (14)$$

where $\bar{u}_\infty$ is defined by

$$\bar{u}_\infty(\bar{x}) = \int_{-\infty}^{\infty} \bar{\gamma}(\bar{\xi} - \bar{x})\bar{p}_{mech}(\bar{\xi})d\bar{\xi}, \qquad (15)$$

and $\bar{\lambda}_{1,2}(\pm 1)$ are the jumps across the joints at $\bar{x} = \pm 1$ of $-(d^3/d\bar{x}^3)\bar{u}$ and $-(d^2/d\bar{x}^2)\bar{u}$, defined by

$$\bar{\lambda}_1(\bar{x}) = \lim_{\Delta\downarrow0} \left([-\frac{d^3\bar{u}}{d\bar{x}^3}(\bar{x} - \Delta) + \frac{d^3\bar{u}}{d\bar{x}^3}(\bar{x} + \Delta)] \mathrm{sgn}(\bar{x}) \right), \qquad (16a)$$

$$\bar{\lambda}_2(\bar{x}) = \lim_{\Delta\downarrow0} \left([-\frac{d^2\bar{u}}{d\bar{x}^2}(\bar{x} - \Delta) + \frac{d^2\bar{u}}{d\bar{x}^2}(\bar{x} + \Delta)] \mathrm{sgn}(\bar{x}) \right). \qquad (16b)$$

$\bar{\lambda}_1(\pm1)$ and $\bar{\lambda}_2(\pm1)$ are the nondimensional layer potentials for the forces and moments exerted on the plate by the joints. They are implicitly determined by the boundary conditions (7) expressing vanishing displacement and slope of the plate at the joints. Differentiating (14) with respect to $\bar{x}$ yields

$$\bar{u}'(\bar{x}) = \bar{u}'_\infty(\bar{x}) - \bar{\lambda}_1(1)\bar{\gamma}'(1 - \bar{x}) + \bar{\lambda}_1(-1)\bar{\gamma}'(-1 - \bar{x})$$

$$+ \bar{\lambda}_2(1)\bar{\gamma}''(1 - \bar{x}) - \bar{\lambda}_2(-1)\bar{\gamma}''(-1 - \bar{x}). \qquad (17)$$

Application of the boundary conditions (7) to expressions (14) and (17) generates the following system of linear equations

$$\begin{pmatrix} -\bar{\gamma}(0) & \bar{\gamma}(2) & \bar{\gamma}'(0) & \bar{\gamma}'(2) \\ -\bar{\gamma}(2) & \bar{\gamma}(0) & \bar{\gamma}'(2) & -\bar{\gamma}'(0) \\ \bar{\gamma}'(0) & \bar{\gamma}'(2) & -\bar{\gamma}''(0) & \bar{\gamma}''(2) \\ \bar{\gamma}'(2) & -\bar{\gamma}'(0) & -\bar{\gamma}''(2) & \bar{\gamma}''(0) \end{pmatrix} \cdot \begin{pmatrix} \bar{\lambda}_1(1) \\ \bar{\lambda}_1(-1) \\ \bar{\lambda}_2(1) \\ \bar{\lambda}_2(-1) \end{pmatrix} = \begin{pmatrix} \bar{u}^s_\infty(1) \\ \bar{u}^s_\infty(1) \\ \bar{u}^{s\prime}_\infty(1) \\ -\bar{u}^{s\prime}_\infty(1) \end{pmatrix} + \begin{pmatrix} \bar{u}^a_\infty(1) \\ -\bar{u}^a_\infty(1) \\ \bar{u}^{a\prime}_\infty(1) \\ \bar{u}^{a\prime}_\infty(1) \end{pmatrix}$$

$$(18)$$

where $\bar{u}_\infty(\bar{x})$ has been split into a symmetric and an anti-symmetric part according to $\bar{u}_\infty(\bar{x}) = \bar{u}^s_\infty(\bar{x}) + \bar{u}^a_\infty(\bar{x})$, where $\bar{u}^s_\infty(-\bar{x}) = \bar{u}^s_\infty(\bar{x})$ and $\bar{u}^a_\infty(-\bar{x}) = -\bar{u}^a_\infty(\bar{x})$.
It is noted that $\bar{\gamma}'(0) = 0$.
The system (18) actually represents the coupled boundary integral equations for the layer potentials $\bar{\lambda}_{1,2}(\pm1)$. However, for the two-dimensional geometry studied here, the integral equations degenerate to algebraic equations, which are readily solved. The solution reads

$$\bar{\lambda}^s_1(1) = \frac{1}{\Delta_1} \left\{ \bar{u}^s_\infty(1)[\bar{\gamma}''(0) - \bar{\gamma}''(2)] + \bar{u}^{s\prime}_\infty(1)\bar{\gamma}'(2) \right\}, \qquad (19a)$$

$$\bar{\lambda}_1^s(-1) = -\bar{\lambda}_1^s(1), \tag{19b}$$

$$\bar{\lambda}_2^s(\pm 1) = \frac{1}{\Delta_1}\left\{-\bar{u}_\infty^s(1)\bar{\gamma}'(2) + \bar{u}_\infty^{s\prime}(1)[\bar{\gamma}(0) + \bar{\gamma}(2)]\right\}, \tag{19c}$$

$$\bar{\lambda}_1^a(\pm 1) = \frac{1}{\Delta_2}\left\{-\bar{u}_\infty^{a\prime}(1)\bar{\gamma}'(2) + \bar{u}_\infty^a(1)[\bar{\gamma}''(0) + \bar{\gamma}''(2)]\right\}, \tag{20a}$$

$$\bar{\lambda}_2^a(1) = \frac{1}{\Delta_2}\left\{\bar{u}_\infty^{a\prime}(1)[\bar{\gamma}(0) - \bar{\gamma}(2)] + \bar{u}_\infty^a(1)\bar{\gamma}'(2)\right\}, \tag{20b}$$

$$\bar{\lambda}_2^a(-1) = -\bar{\lambda}_2^a(1), \tag{20c}$$

where $\bar{\lambda}_{1,2}^s(\pm 1)$ is the solution of (18) for symmetric excitation and $\bar{\lambda}_{1,2}^a(\pm 1)$ is the solution of (18) for anti-symmetric excitation. (Note that $\bar{\lambda}_1^s$ is anti-symmetric, while $\bar{\lambda}_1^a$ is symmetric!) The general solution of (18) is given by $\bar{\lambda}_{1,2}(\pm 1) = \bar{\lambda}_{1,2}^s(\pm 1) + \bar{\lambda}_{1,2}^a(\pm 1)$.

The determinant factors Δ_1 and Δ_2 are given by

$$\Delta_1 = [\bar{\gamma}(0) + \bar{\gamma}(2)][-\bar{\gamma}''(0) + \bar{\gamma}''(2)] - [\bar{\gamma}'(2)]^2, \tag{19d}$$

$$\Delta_2 = [-\bar{\gamma}(0) + \bar{\gamma}(2)][\bar{\gamma}''(0) + \bar{\gamma}''(2)] - [\bar{\gamma}'(2)]^2. \tag{20d}$$

The special case of a central line force, i.e. $p_{mech}(x) = F'\delta(x)$ yields:

$$\bar{u}_\infty(\bar{x}) = \bar{u}_\infty^s(\bar{x}) = \bar{\gamma}(-\bar{x}) = \bar{\gamma}(\bar{x}) \text{ and } \bar{u}_\infty^a(\bar{x}) \equiv 0,$$

while excitation by a central line moment, i.e. $p_{mech}(x) = -M'\delta'(x)$ yields (take $F' = M'/a$ for normalization):

$$\bar{u}_\infty(\bar{x}) = \bar{u}_\infty^a(\bar{x}) = \bar{\gamma}'(-\bar{x}) = -\bar{\gamma}'(\bar{x}), \text{ and } \bar{u}_\infty^s(\bar{x}) \equiv 0.$$

Power flow

The input power per unit length delivered by the mechanical force distribution is given by (apply formula (A.4) from Appendix A):

$$\bar{\Pi}_{in}' = \frac{1}{2}\,\text{Re}\left[\int_{-\infty}^{\infty} \bar{p}_{mech}(\bar{x})^* \cdot -i\bar{u}(\bar{x})d\bar{x}\right], \tag{21a}$$

where $\bar{\Pi}'_{in} = \Pi'_{in}/(\omega F'^2 a^3/D)$. For the case of a central line force we have

$$\bar{\Pi}'^s_{in} = \frac{1}{2} \, \text{Im}[\bar{u}(0)], \tag{21b}$$

while for the case of a central line moment we have

$$\bar{\Pi}'^a_{in} = \frac{1}{2} \, \text{Im}\,[\bar{u}'(0)], \tag{21c}$$

where for nondimensionalization purposes F' has been replaced by M'/a.

Farfield pressure and directivity function

It is well known that the farfield pressure is given by the asymptotic form

$$\bar{p}(\bar{r},\theta) \sim (2/\pi\beta\bar{r})^{\frac{1}{2}} \exp[i\beta\bar{r} - \pi/4] \, D(\theta), \quad \beta\bar{r} >> 1, \tag{22a}$$

where $D(\theta)$ is the directivity function given by

$$D(\theta) = \frac{1}{2i} \frac{\varepsilon\beta^5}{M^6} \hat{u}(\kappa), \quad \kappa = \beta\sin\theta, \quad -\pi/2 \le \theta \le \pi/2, \tag{22b}$$

and $\hat{u}(\kappa)$ is the Fourier transform of $\bar{u}(\bar{x})$ defined by

$$\hat{u}(\kappa) = \int_{-\infty}^{\infty} \bar{u}(\bar{x}) \exp(-i\kappa\bar{x}) d\bar{x}. \tag{22c}$$

Polar co-ordinates $(\bar{r},\theta)$, where $\bar{x} = \bar{r}\sin\theta$, $y = \bar{r}\cos\theta$, $-\pi/2 \le \theta \le \pi/2$, have been introduced in relation (22a).

It follows from (14) and (15) as well as the properties of Fourier transforms (Appendix D) that

$$\hat{u}(\kappa) = \Big[\hat{p}_{mech}(\kappa) + \bar{\lambda}_1(1)\exp(-i\kappa) - \bar{\lambda}_1(-1)\exp(i\kappa)$$

$$\tag{23a}$$

$$-\bar{\lambda}_2(1)\cdot -i\kappa\exp(-i\kappa) + \bar{\lambda}_2(-1)\cdot -i\kappa\exp(i\kappa)\Big]\hat{\gamma}(\kappa),$$

where $\hat{p}_{mech}(\kappa)$ and $\hat{\gamma}(\kappa)$ are the Fourier transforms of $\bar{p}_{mech}(\bar{x})$ and $\bar{\gamma}(\bar{x})$ respectively. The symmetric case with line drive at $\bar{x} = 0$ yields

$$\hat{u}^s(\kappa) = \Big[1 + 2\bar{\lambda}_1^s(1)\cos(\kappa) + 2\bar{\lambda}_2^s(1)\kappa\sin(\kappa)\Big]\hat{\gamma}(\kappa), \tag{23b}$$

while the antisymmetric case (i.e. moment excitation at $\bar{x} = 0$) yields

$$\hat{u}^a(\kappa) = \left[-i\kappa - 2i\bar{\lambda}_1^a(1)\sin(\kappa) + 2i\bar{\lambda}_2^a(1)\kappa\cos(\kappa)\right]\hat{\gamma}(\kappa). \qquad (23c)$$

The Fourier transform of $\bar{\gamma}(\bar{x})$ follows from expression (C.13) in Appendix C. The result is

$$\hat{\gamma}(\kappa) = \left[(\kappa^4 - \alpha^4) - i\frac{\varepsilon}{M}\alpha^5(\beta^2 - \kappa^2)^{-\frac{1}{2}}\right]^{-1}, \qquad (24)$$

where $\alpha = \beta/M$. Note the difference in nondimensionalization between relation (24) and Appendix C. After some manipulations one has

$$\hat{\gamma}(\kappa = \beta\sin\theta) = \frac{M^6}{\beta^4}\frac{\cos\theta}{(M^4\sin^4\theta - 1)M^2\cos\theta - i\varepsilon}. \qquad (25)$$

At low frequencies, i.e. for $\beta << 1 \Rightarrow \kappa << 1$, the line-force excited plate radiates like a dipole in normal direction, since its directivity is proportional to

$$\hat{u}^s(\kappa) \sim \left[1 + 2\bar{\lambda}_1^s(1) + O(\kappa^2)\right]\frac{M^6}{\beta^4}\frac{\cos\theta}{-i\varepsilon}, \qquad (26a)$$

while for the moment-driven plate a quadrupole directivity is found:

$$\hat{u}^a(\kappa) \sim \left[1 + 2i\bar{\lambda}_1^a(1) - 2\bar{\lambda}_2^a(1) + O(\kappa^2)\right]\frac{M^6}{\beta^3}\frac{\sin(2\theta)}{2\varepsilon}. \qquad (26b)$$

3   Numerical results and discussion

The formulas presented in the preceeding sections have been used to obtain numerical results for the input power, the plate response and the farfield directivity for a steelplate in water characterized by the following values for the physical and geometrical quantities:

$$c_0 = 1.49 \times 10^3 \, m/s, \quad \rho_0 = 10^3 \, kg/m^3, \quad E = 0.216 \times 10^{12} \, N/m^2,$$

$$\rho_p = 7.84 \times 10^3 \, kg/m^3, \quad \nu = 0.276, \quad h = 0.05 \, m, \quad a = 0.5 \, m.$$

The plate has been excited by a central line force and a central line moment respectively.

Input power and resonance frequencies

First, the input power delivered by the mechanical drive was calculated from expressions (21b) and (21c) for a wide range of frequencies. Inspection of the computed spectral data immediately brings up the resonance frequencies for the fluid-loaded plate. A close view at the power spectrum for the symmetrically driven plate is given in Figure 3.

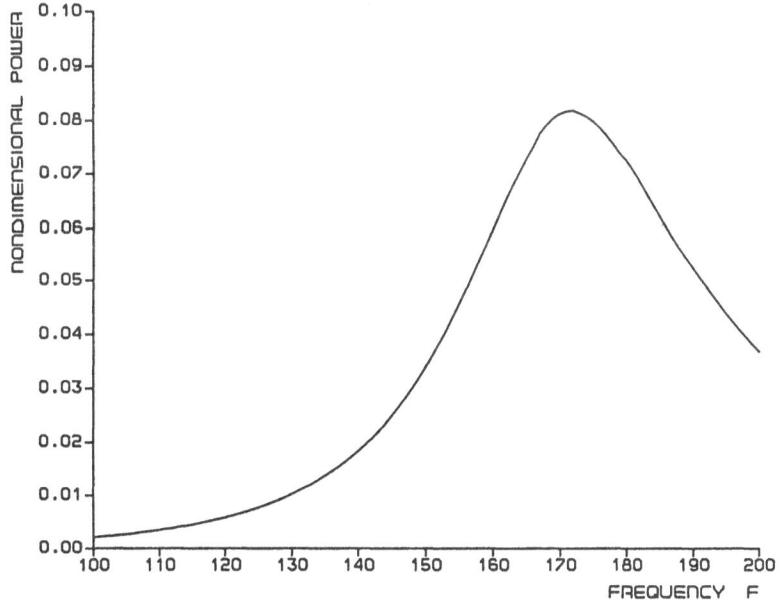

FIGURE 3. Nondimensional input power per unit length vs. frequency (in Hz) for line-force excited, fluid-loaded strip

The results clearly show the maximum of the power flow at 172 Hz, which is the first fluid-loaded resonance frequency. The corresponding resonance for a clamped strip vibrating in vacuo appears at 281 Hz. This value is readily computed from the roots of the frequency equation

$$\tan(k_p a) = \pm \tanh(k_p a),$$

where the minus (plus) sign corresponds to (anti-)symmetric vibrations. Unlike the in vacuo behaviour the plate vibration is damped due to radiation losses and consequently, the amplitude of the plate response is finite on passage through resonance.

By inspecting the power spectrum for both the symmetric and the antisymmetric case for frequencies up to coincidence (4482.7 Hz), the full set of resonances was identified. The results are summarized in Table I.

TABLE I. Fluid-loaded, resonance frequencies (Hz) estimated from spectral data for the input power into the strip, compared to the in vacuo resonances. Odd (even) modenumbers n correspond to (anti-)symmetric vibrations. Accuracy of $f_n^{fl.l}$ is 1 Hz for $n = 1$ and 5 Hz for $n \geq 2$.

n	1	2	3	4	5	6
f_n^{vac}	280.7	774.6	1516.7	2509.7	3773.9	5236.3
$f_n^{fl.l}$	172	610	1310	2220	3310	4690

All fluid-loaded resonance frequencies are shifted downwards relative to the in vacuo natural frequencies. This effect may be attributed mainly to the added mass of the fluid volume that is accelerated and decelerated by the plate's vibrational motion.

Furthermore, it is interesting to note that the influence of fluid-loading on the relative shift of the resonance frequencies decreases with increasing modenumbers.

Plate response at resonance

The displacement field of the strip has been calculated from expression (14) for each of the resonance frequencies. The results are presented in Figure 4.

For the fundamental mode (170 Hz) the results look quite the same as one would expect for the plate without fluid-loading. The strip vibrates in phase, i.e. without a nodal pattern. Note however, that the displacement is not in phase with the exciting force, as would be the case in vacuo.

Although not shown in the figures the phase of the plate response makes a jump of 180° across the joints.

A very similar interpretation might be given for the shape of the first anti-symmetric mode (610 Hz), except that a node appears at the origin where the line moment acts, while two antinodes are located symmetrically around the origin.

Figure 4c shows the displacement field of the third mode. There are apparently two distinct regions of the strip that vibrate roughly with opposite phase, just as one would expect. However, the phase is not uniform as for a pure standing wave, but there is an identifiable slope, i.e. part of the displacement field is a one-way wave travelling towards infinity, while the main part is a standing wave.

The absence of a pure nodal point is an even more striking feature of the response that makes it different from the in vacuo case. Although there is a distinct point of minimum amplitude at $\bar{x} = 0.28$, the amplitude of the vibration at this point still attains a value of about 18% of the amplitude at the drive.

These unusual observations for both amplitude and phase might be given a physical explanation in terms of fluid-loading. Since part of the wavefield on a fluid-loaded plate is damped by acoustic radiation this part looses energy and therefore its amplitude decreases during propagation. Upon arrival at the joint ($\bar{x} = 1$) the flexural wave is reflected and starts propagating in opposite direction, while radiation continues. Moreover, the subsonic surface wave component of the acoustic field that was travelling along with the flexural wave (without attenuation!) is dumped across the joint without being reflected. As a consequence, the reflection process produces additional radiation losses.

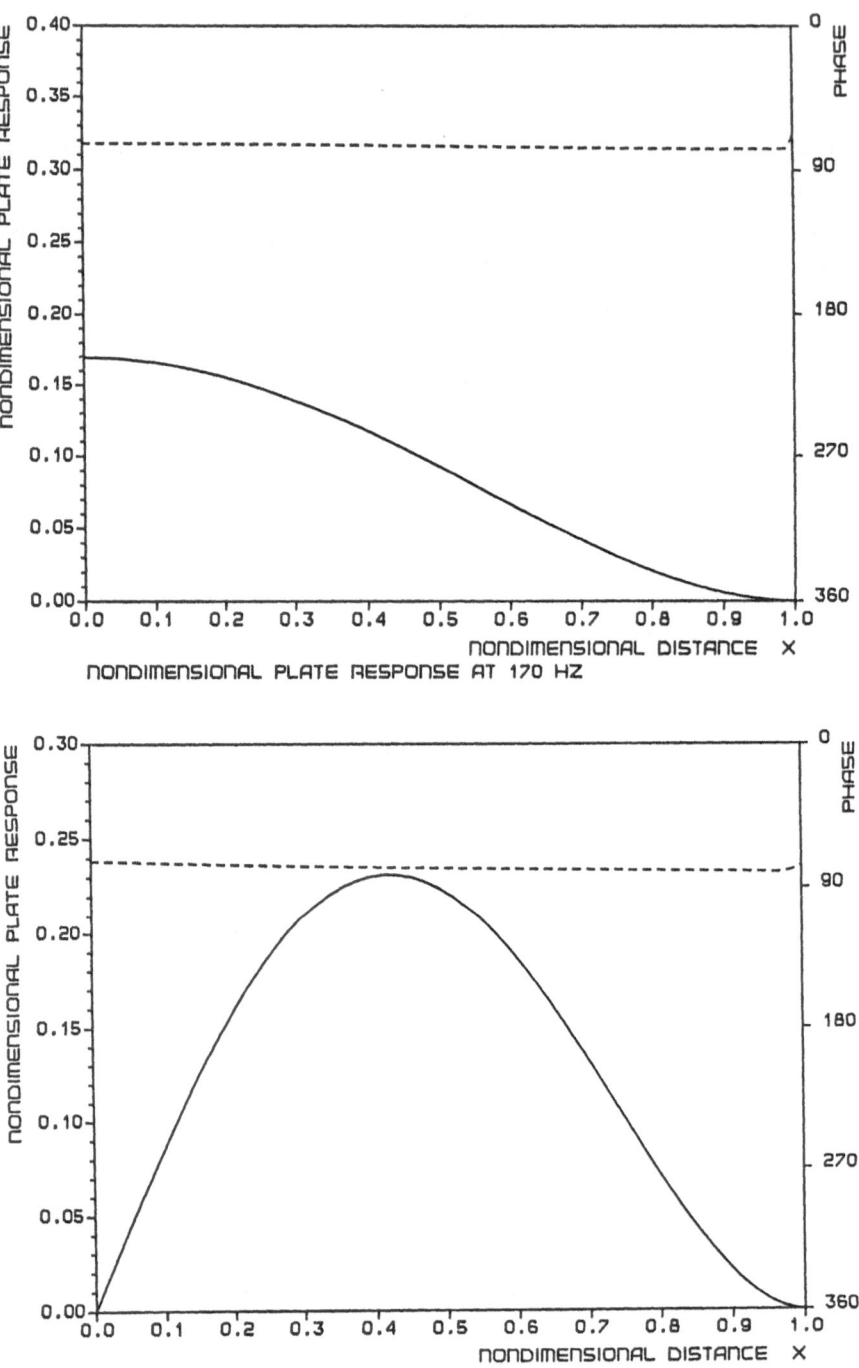

FIGURE 4. Plate response for line–driven, fluid–loaded strip. Solid line: modulus, dashed line: phase (in degrees, relative to the drive).   (a) 170 Hz; (b) 610 Hz;

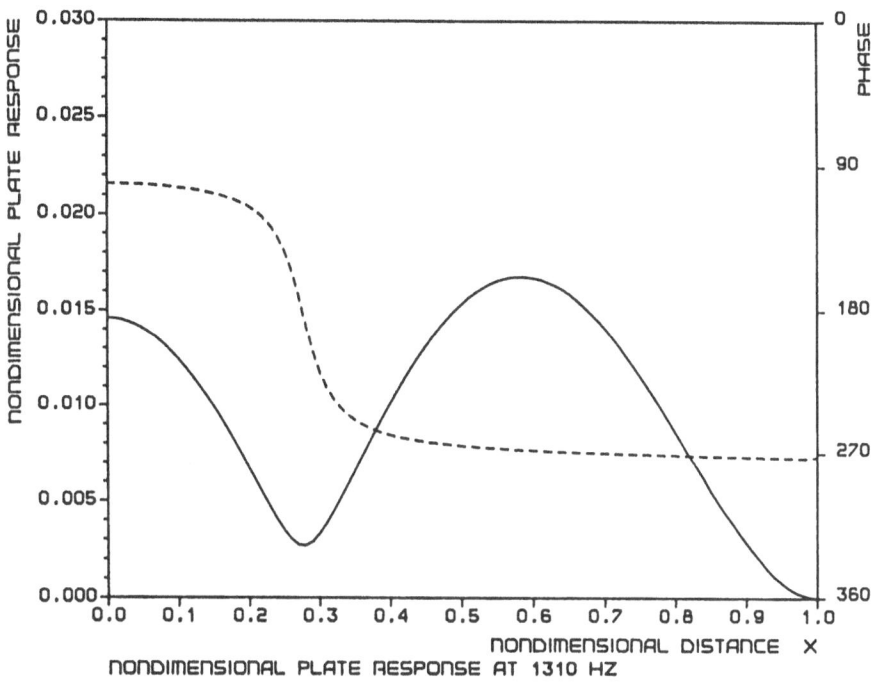

NONDIMENSIONAL PLATE RESPONSE AT 1310 HZ

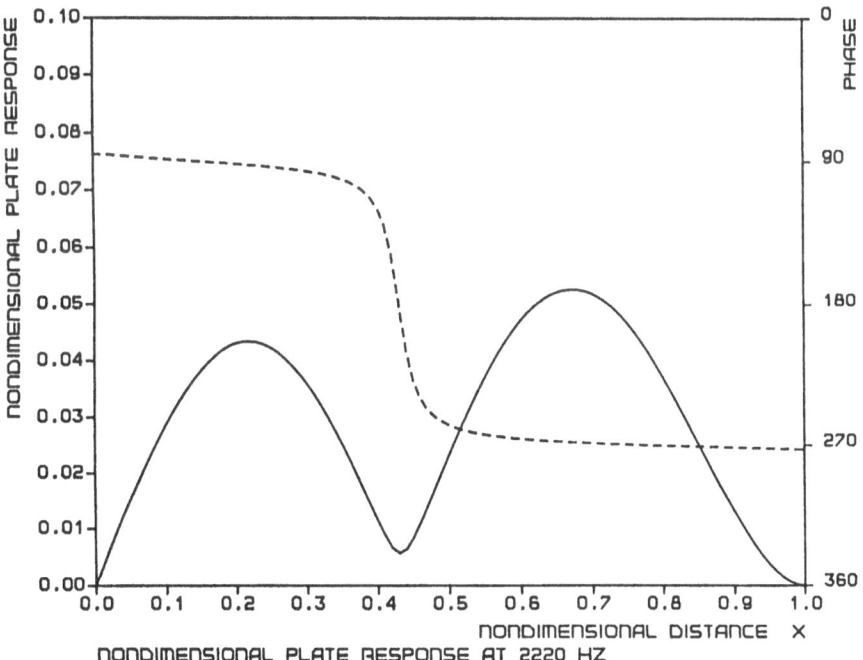

NONDIMENSIONAL PLATE RESPONSE AT 2220 HZ

FIGURE 4. (Continued) (c) 1310 Hz; (d) 2220 Hz;

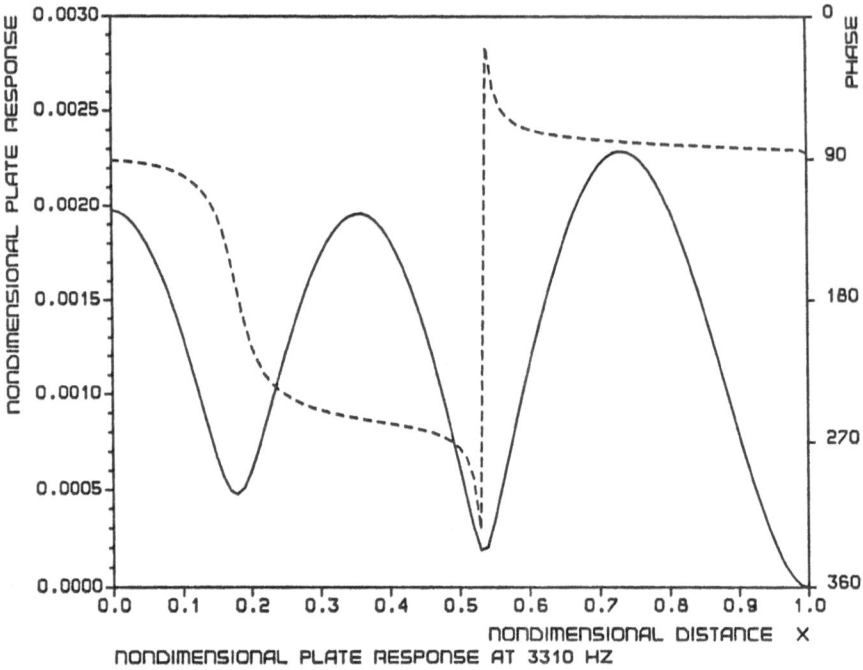

NONDIMENSIONAL PLATE RESPONSE AT 3310 HZ

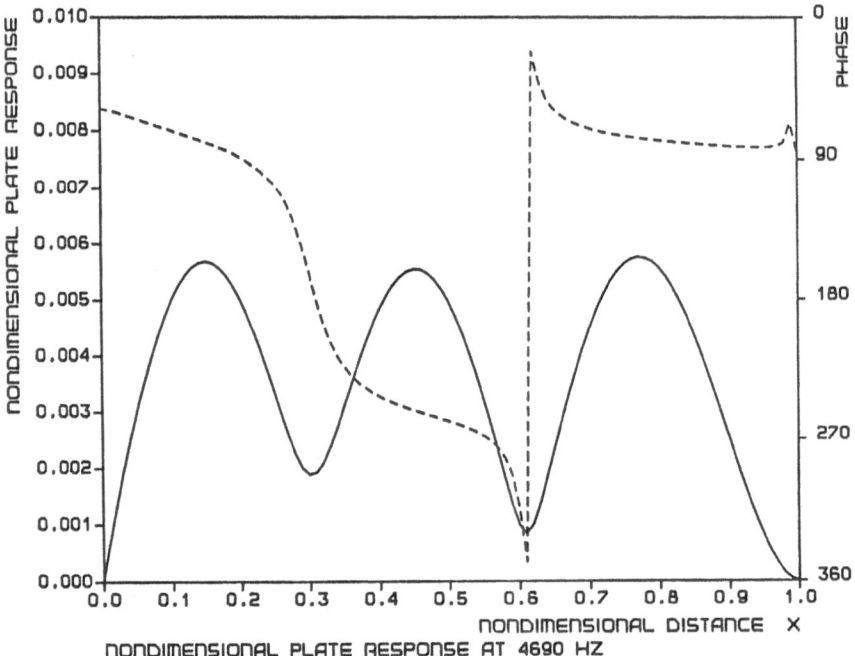

NONDIMENSIONAL PLATE RESPONSE AT 4690 HZ

FIGURE 4. (Continued) (e) 3310 Hz; (f) 4690 Hz.

Now, the slope of the phase of the displacement field that indicates the existence of a propagating, one-way wave can actually be attributed to a wave that is damped by radiation losses. Similarly, because the damping process reduces the amplitude of the plate wave during propagation and reflection, the reflected wave is unable to compensate completely the amplitude of the 'incoming' wave at those points where the phases of the waves differ by 180°.

A last remarkable feature in Figure 4c is that the value of the plate response at the antinodal points ($\bar{x} = \pm 0.58$) is about 10% higher than at the central antinode ($\bar{x} = 0$). This effect might be attributed to the difference in compliance between the coplanar, semi-infinite elastic plate (for $\bar{x} > 1$) and the remainder of the strip ($-1 < \bar{x} < 0.28$).

The effects that have been observed in Figure 4c are also present in the figures that show the resonant vibrations for higher order modes.

Additionally, it is noted that the formation of nodal points close to the excitation is more poor than for points close to the joints. This observation reconfirms the explanation given here for the appearance of such 'quasi-nodes'. The wave that should cancel approximately the incoming wave at the inner node has to travel a distance (to and from the reflecting joint) which is about twice as long as for the outer node. Therefore, its amplitude has been diminished further and consequently the failure to compensate the incoming wave at the foreseen nodal point is more pronounced.

Finally, in Figure 4f the situation is more complex because this resonance appears above coincidence. Therefore, radiation dominates over the compliance of the coplanar plates.

Directivity function at resonance

Polar plots for the directivity function are given in Figure 5 for the resonance frequencies.

A general feature encountered in these figures is that the pressure goes to zero at grazing angles. This is a consequence of the compliance of the coplanar, semi-infinite plates.

As a general rule one can state that the number of directive lobes increases with increasing modenumber. It is also clear from Figures 5c

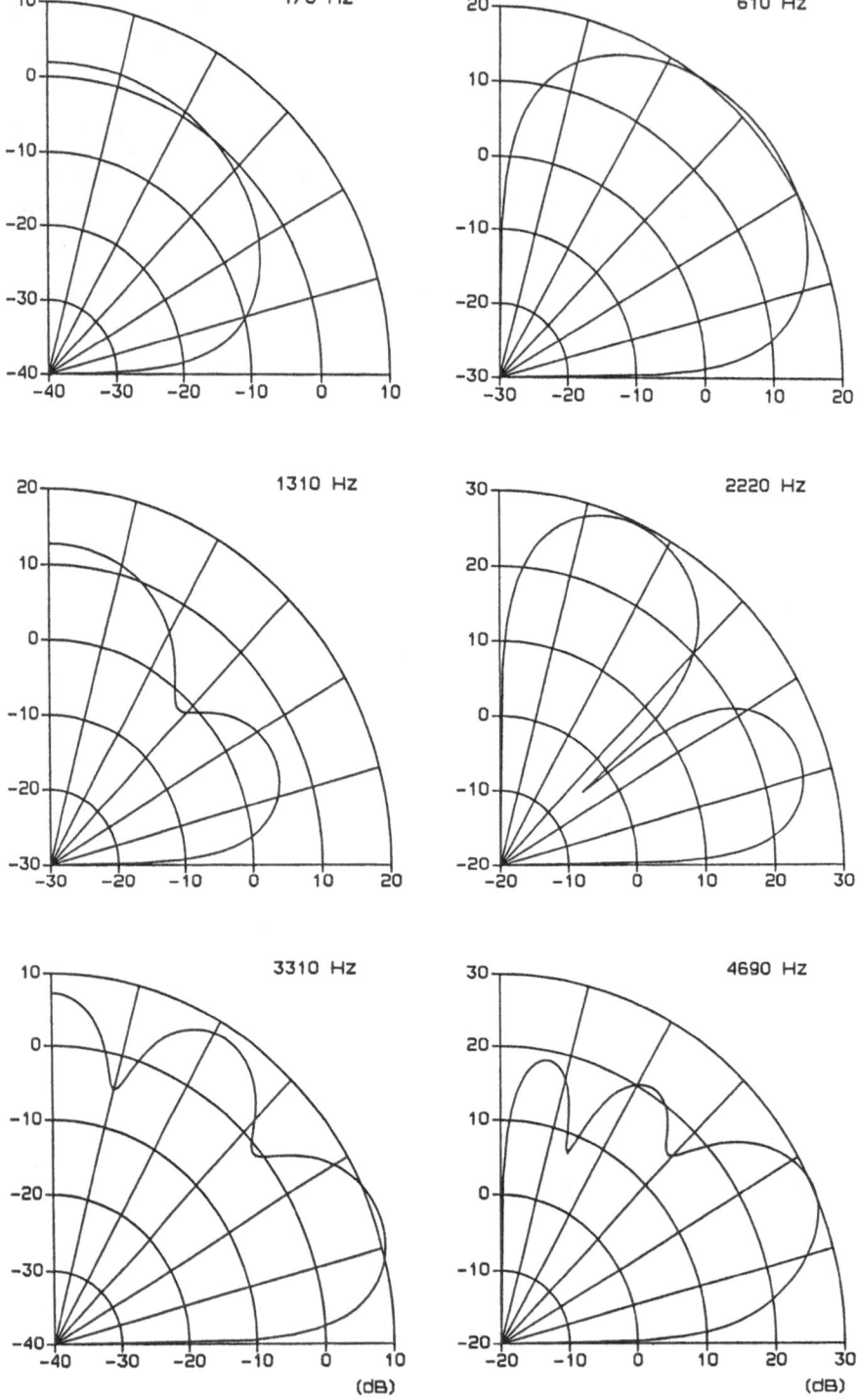

FIGURE 5. Polar diagram for $20 \log_{10} |D(\theta)|$.

(a) 170 Hz; (b) 610 Hz; (c) 1310 Hz; (d) 2220 Hz; (e) 3310 Hz; (f) 4690 Hz.

and 5d that the directive lobes corresponding to the outer ventral regions
of the plate have a smaller amplitude than the inner ones. This means
that the periferal strip regions do not radiate sound as effectively as the
inner ones. This observation reconfirms the compliance argument given
before.

Influence of thickness to width ratio on fundamental frequency

This section addresses a problem which is of practical importance to
designers of ships.
One of the methods to reduce annoyance due to noise and vibration
on board of ships is to choose the dimensions of the structural elements in such a way that resonant vibrations are unlikely to occur. So
the resonance frequencies should be outside the range of frequencies at
which the engines and the propeller usually operate. This method of
passive noise control is especially important for the fundamental resonance, since high frequency noise and vibration levels are easily reduced
by the application of sound absorbing materials and resilient mounting
systems, while these measures fail at low frequencies.
So, the interest lies in the variation of the first fluid-loaded resonance
frequency with variations in plate thickness and width.
The in vacuo natural frequencies depend on a and h according to

$$
f_n^{vac} = \frac{\lambda_n^2}{2\pi} \frac{h}{a^2} \left(\frac{E/\rho_p}{12(1 - \nu^2)} \right)^{\frac{1}{2}},
$$

where λ_n are the roots of the frequency equation $\tan \lambda_n = \pm \tanh \lambda_n$.
Therefore, the results presented here were obtained by keeping a fixed,
while h was varied and both f_1^{vac} and $f_1^{fl.l}$ are calculated. The results
are presented in Table II.
The results indicate that the relative shift of the fundamental fluid-
loaded resonance frequency increases with decreasing thickness. The
physical explanation of this effect is based on the influence of the added
mass of the fluid which becomes more important if the plate thickness
— and thus its mass per unit area — is small.

TABLE II. Variation of first, fluid-loaded resonance frequency with thickness.

$h(mm)$	h/a	$f_1^{fl.l}(Hz)$	$f_1^{fl.l}/f_1^{vac}$
50	0.1	172 ± 1	0.613
40	0.08	128 ± 1	0.570
25	0.05	69 ± 1	0.492
20	0.04	51 ± 1	0.454
10	0.02	19.5 ± 0.5	0.347
5	0.01	7.25 ± 0.25	0.258
2.5	0.005	2.6 ± 0.1	0.185
1	0.002	0.66 ± 0.02	0.118
0.5	0.001	0.24 ± 0.01	0.0855

Acknowledgement

This work is supported by the TNO Institute of Applied Physics, Delft, The Netherlands.

References

General introductory literature on acoustics and structural vibrations:

1. J.W. Strutt Lord Rayleigh, *The Theory of Sound*, (Dover, New York, 1945), second edition, Vols. I and II.

2. P.M. Morse and K.U. Ingard, *Theoretical Acoustics*, (McGraw-Hill, New York, 1968, reprint: Princeton University Press, 1986).

3. L. Cremer and M. Heckl, *Structure-Borne Sound*, (Berlin, Springer, 1988), second edition.

4. A.D. Pierce, *Acoustics: an introduction to its physical principles and applications*, (McGraw-Hill, New York, 1981, reprint: Acoustical Society of America / American Institute of Physics, 1989).

Specific literature on fluid-loaded plates and shells:

5. M.C. Junger and D. Feit, *Sound, Structures and Their Interaction*, (Cambridge MA, MIT-press, 1986), second edition.

6. D.G. Crighton, *The 1988 Rayleigh medal lecture: fluid-loading — the interaction between sound and vibration*, J. Sound Vib. **133**, 1-27 (1989).

7. P.R. Nayak, *Line admittance of infinite isotropic fluid-loaded plates*, J. Acoust. Soc. Am. **47**, 191-201 (1970).

8. D.G. Crighton, *The free and forced waves on a fluid-loaded elastic plate*, J. Sound Vib. **62**, 225-235 (1979).

9. D. Feit and Y.N. Liu, *The nearfield response of a line-driven fluid-loaded plate*, J. Acoust. Soc. Am. **78**, 763-766 (1985).

Appendices

A   Complex arithmetics for simple time-harmonic quantities

If the real-valued, physical quantity $q(\boldsymbol{x}, t)$ has a simple time-harmonic dependence (circular frequency ω) it may be written in the form

$$q(\boldsymbol{x}, t) = q_c(\boldsymbol{x}) \cos \omega t - q_s(\boldsymbol{x}) \sin \omega t, \tag{A.1}$$

where the functions q_c and q_s are unique.

By definition, a complex-valued function $q(\boldsymbol{x}; \omega)$ is associated to $q(\boldsymbol{x}, t)$ according to

$$q(\boldsymbol{x}; \omega) = q_c(\boldsymbol{x}) - i q_s(\boldsymbol{x}), \tag{A.2}$$

i being the imaginary unit. It is easy to verify that $q(\boldsymbol{x}, t)$ is related to $q(\boldsymbol{x}; \omega)$ by

$$q(\boldsymbol{x}, t) = \mathrm{Re}[q(\boldsymbol{x}; \omega) \exp(-i\omega t)]. \tag{A.3}$$

For notational reasons only it is very practical to work with $q(\boldsymbol{x}; \omega)$ instead of both $q_c(\boldsymbol{x})$ and $q_s(\boldsymbol{x})$. So the operator Re is dropped as well as the time factor $\exp(-i\omega t)$. However, it is noted that taking the time-derivative of a real-valued physical quantity corresponds to multiplying the associated complex counterpart by a factor $-i\omega$.

A useful formula concerns the time-average of a product of two time-harmonic functions:

$$\langle p(\boldsymbol{x}, t) q(\boldsymbol{x}, t) \rangle_T \equiv \lim_{T \to \infty} \frac{1}{2T} \int_{-T}^{T} p(\boldsymbol{x}, t) q(\boldsymbol{x}, t) dt = \frac{1}{2} \mathrm{Re}[p(\boldsymbol{x}; \omega) q(\boldsymbol{x}; \omega)^*], \tag{A.4}$$

where q^* denotes the complex conjugate of q. This lemma is easily verified by inserting (A.1) and (A.2) in the left-hand-side and the right-hand-side of (A.4) respectively.

Finally, two remarks on the notations used in the main text.

First, the argument ω appearing in $q(\boldsymbol{x}; \omega)$ has been dropped throughout for brevity. Second, since there is nothing ambiguous in the notation both $q(\boldsymbol{x}, t)$ and $q(\boldsymbol{x}; \omega)$ are indicated by the symbol "q".

B   Green integral representation for the acoustic pressure

In this appendix the Neumann problem for the Helmholtz equation in the halfspace $\bar{z} > 0$, (problem (8), (9)) is solved by using the appropriate Green function. This function satisfies

$$\bar{\Delta}\bar{G} + \beta^2 \bar{G}(\bar{x}, \bar{z}; \bar{\xi}, \bar{\zeta}) = \delta(\bar{x} - \bar{\xi}, \bar{z} - \bar{\zeta}), \quad -\infty < \bar{x} < \infty, \quad \bar{z} > 0, \quad \text{(B.1)}$$

$$\frac{\partial \bar{G}}{\partial \bar{z}}(\bar{x}, 0; \bar{\xi}, \bar{\zeta}) = 0, \quad -\infty < \bar{x} < \infty, \quad \bar{z} = 0, \quad \text{(B.2)}$$

and a radiation condition at infinity. The source $(\bar{\xi}, \bar{\zeta})$ is located in the halfspace $\bar{z} > 0$, i.e. $\bar{\zeta} \geq 0$. A solution of (B.1) that satisfies the radiation condition is $-\frac{i}{4}H_0^{(1)}(\beta \bar{R}_+)$, where $\bar{R}_+ = \sqrt{(\bar{x} - \bar{\xi})^2 + (\bar{z} - \bar{\zeta})^2}$, so the Green function can be constructed in the form

$$\bar{G}(\bar{x}, \bar{z}; \bar{\xi}, \bar{\zeta}) = -\frac{i}{4}H_0^{(1)}(\beta \bar{R}_+) + \bar{v}(\bar{x}, \bar{z}; \bar{\xi}, \bar{\zeta}), \quad \text{(B.3)}$$

where $\bar{v}(\bar{x}, \bar{z}; \bar{\xi}, \bar{\zeta})$ satisfies the homogeneous Helmholtz equation, a radiation condition and the Neumann condition

$$\frac{\partial \bar{v}}{\partial \bar{z}}(\bar{x}, 0; \bar{\xi}, \bar{\zeta}) = -\frac{\partial}{\partial z}[-\frac{i}{4}H_0^{(1)}(\beta \bar{R}_+)]_{\bar{z}=0} = -\frac{i}{4}H_1^{(1)}(\beta \bar{R}_+|_{\bar{z}=0})\beta \frac{-\bar{\zeta}}{\bar{R}_+|_{\bar{z}=0}}.$$
$$\text{(B.4)}$$

The function $\bar{v}(\bar{x}, \bar{z}; \bar{\xi}, \bar{\zeta})$ is easily obtained by taking the image with respect to the plane $\bar{z} = 0$ of the point source solution in $(\bar{\xi}, \bar{\zeta})$. The result reads:

$$\bar{v}(\bar{x}, \bar{z}; \bar{\xi}, \bar{\zeta}) = -\frac{i}{4}H_0^{(1)}(\beta \bar{R}_-), \quad \text{where } \bar{R}_- = \sqrt{(\bar{x} - \bar{\xi})^2 + (\bar{z} + \bar{\zeta})^2}. \quad \text{(B.5)}$$

Finally, application of Green's theorem to $\bar{p}\bar{\Delta}\bar{G} - \bar{G}\bar{\Delta}\bar{p}$ yields the integral representation

$$\bar{p}(\bar{x}, \bar{z}) = \sigma^4 \int_{-\infty}^{\infty} \bar{G}(\bar{x}, \bar{z}; \bar{\xi}, \bar{\zeta} = 0)\bar{u}(\bar{\xi})d\bar{\xi} \quad \text{(B.6)}$$

where
$$\bar{G}(\bar{x}, \bar{z}; \bar{\xi}, \bar{\zeta} = 0) = \frac{1}{2i}H_0^{(1)}(\beta \bar{R}_0), \quad \text{and } \bar{R}_0 = \bar{R}_\pm|_{\bar{\zeta}=0} = \sqrt{(\bar{x} - \bar{\xi})^2 + \bar{z}^2}.$$

C　The fluid-loaded, infinite plate Green kernel

In this appendix the fundamental solution for the line-driven, fluid-loaded, infinite plate is discussed in some detail.

First, it is noted that different scalings than those used in the text will be more appropriate to the present problem. To be specific, the plate response will be made nondimensional by the in vacuo response at the drive, i.e. $(-i\omega)^{-1}k_p F_0'/4m\omega$ while the co-ordinate along the plate will be scaled by the flexural wavelength k_p^{-1}. These scalings will show up naturally from the analysis. By introducing these scalings it will become clear that the Green kernel depends on two nondimensional parameters only, i.e. ε and M.

The governing equations for the line-driven, fluid-loaded, infinite plate are

$$D\frac{d^4u}{dx^4} - m\omega^2 u(x) = F_0'\delta(x - x_0) - p(x,0), \quad -\infty < x < \infty, \quad (C.1)$$

$$\{\Delta + k_0^2\}p(x,z) = 0, \quad -\infty < x < \infty, \quad z > 0, \quad (C.2)$$

$$\frac{\partial p}{\partial z}(x,0) = \rho_0\omega^2 u(x), \quad -\infty < x < \infty, \quad z = 0, \quad (C.3)$$

where $m = \rho_p h$ is the mass per unit area and $k_0 = \omega/c_0$ is the acoustic wavenumber.

Equations (C.1-3) are readily solved by means of Fourier integral transforms. Fourier integral transform pairs are defined by

$$\tilde{u}(k_x) = \int_{-\infty}^{\infty} u(x)\exp(-ik_x x)dx, \quad (C.4)$$

$$u(x) = \frac{1}{2\pi}\int_C \tilde{u}(k_x)\exp(ik_x x)dk_x, \quad (C.5)$$

$$\tilde{p}(k_x;z) = \int_{-\infty}^{\infty} p(x,z)\exp(-ik_x x)dx, \quad (C.6)$$

$$p(x,z) = \frac{1}{2\pi}\int_C \tilde{p}(k_x;z)\exp(ik_x x)dk_x, \quad (C.7)$$

where the contour C runs along the real axis in the complex k_x-plane. Poles located on this contour are accounted for by taking the Cauchy principal value for the integral, while the radiation condition determines whether the associated semi-circular paths around these poles should be taken in clockwise or anticlockwise direction.

Taking the Fourier transform of equations (C.1-3) yields

$$(Dk_x^4 - m\omega^2)\tilde{u}(k_x) = F_0' \exp(-ik_x x_0) - \tilde{p}(k_x; 0), \qquad (C.8)$$

$$\left[\frac{\partial^2}{\partial z^2} + (k_0^2 - k_x^2)\right]\tilde{p}(k_x; z) = 0, \quad z > 0, \qquad (C.9)$$

$$\frac{d}{dz}[\tilde{p}(k_x; 0)] = \rho_0\omega^2\tilde{u}(k_x), \quad z = 0. \qquad (C.10)$$

The solution of (C.9) that satisfies the radiation condition is

$$\tilde{p}(k_x; z) = \tilde{p}(k_x; 0) \exp[ik_z z], \qquad (C.11)$$

where the branch of the two-valued function $k_z = (k_0^2 - k_x^2)^{\frac{1}{2}}$ is chosen in such a way that for real values of k_x :

$$k_z = \begin{cases} |k_0^2 - k_x^2|^{\frac{1}{2}}, & -k_0 < k_x < k_0, \\ i\,|k_x^2 - k_0^2|^{\frac{1}{2}}, & |k_x| > k_0. \end{cases}$$

From (C.10) and (C.11) it follows that

$$ik_z\tilde{p}(k_x; 0) = \rho_0\omega^2\tilde{u}(k_x). \qquad (C.12)$$

Solving the equations (C.8), (C.11) and (C.12) for $\tilde{u}(k_x)$, $\tilde{p}(k_x; 0)$ and $\tilde{p}(k_x; z)$ yields:

$$\tilde{u}(k_x) = \frac{(-i\omega)^{-1}F_0' \exp(-ik_x x_0)}{Z_p(k_x) + Z_a(k_x)}, \qquad (C.13)$$

$$\tilde{p}(k_x; 0) = \frac{Z_a(k_x)F_0' \exp(-ik_x x_0)}{Z_p(k_x) + Z_a(k_x)}, \qquad (C.14)$$

where $Z_p(k_x)$ and $Z_a(k_x)$ are given by

$$Z_p(k_x) = (-i\omega)^{-1}(Dk_x^4 - m\omega^2), \quad Z_a(k_x) = \rho_0\omega/k_z.$$

The evaluation of the inverse transforms should be carried out in the complex k_x-plane.

Either direct integration along the contour C or indirect integration by means of the residue theorem may be used. The latter method avoids numerical evaluation of Cauchy principle value integrals. However, the roots of the dispersion relation, i.e.

$$Z_p(k_x) + Z_a(k_x) = 0, \qquad (C.15)$$

must be known and an integral along the branch cut of k_z appears. Except for the numerical calculation of these roots a closed form integral expression for $u(x)$ is obtained, which has to be evaluated numerically. The roots of the dispersion relation are classified as follows. Relation (C.15) may be rationalized upon multiplying by the complementary relation
$Z_p(k_x) - Z_a(k_x) = 0$. After some manipulations a fifth degree polynomial equation in k_x^2 is obtained in the following nondimensional form:

$$(\bar{k}_x^4 - 1)^2(\bar{k}_x^2 - M^2) - \varepsilon^2/M^2 = 0, \qquad (C.16)$$

where $\bar{k}_x$, ε and M are given by

$$\bar{k}_x = k_x/k_p, \quad \varepsilon = \frac{\rho_0 c_0}{\rho_p h \omega_c}, \quad M = k_0/k_p = \left(\frac{\omega}{\omega_c}\right)^{\frac{1}{2}},$$

while k_p is the flexural wavenumber given by $k_p^4 = m\omega^2/D$ and ω_c is the coincidence frequency defined by $\omega_c = (c_0^2/h)(E/12(1 - \nu^2)\rho_p)^{-\frac{1}{2}}$.
From (C.16) it is clear that the roots are determined by two nondimensional parameters only, i.e. ε and M. Equation (C.16) has five roots for $\bar{k}_x^2$, the complex ones appearing in conjugate pairs, so at least one of them is real. However, the interest is in the roots for $\bar{k}_x$ of the nondimensional version of (C.15), i.e.

$$(\bar{k}_x^4 - 1) - \frac{\varepsilon}{M}i(M^2 - \bar{k}_x^2)^{-\frac{1}{2}} = 0, \qquad (C.17)$$

so only five of the ten roots of (C.16) are roots of (C.17), the remaining five are spurious, since they correspond to the complementary dispersion relation.
Because u is symmetric with respect to x_0, the inverse Fourier integral for $u(x)$ is evaluated for $x \geq x_0$ only, while x_0 is taken to be zero without

loss of generality.

The contour C is closed by a semi-circle at infinity in the upper halfplane to meet the radiation and boundedness conditions. Furthermore, if the contour C is slightly displaced from the real axis to include poles located on the positive part of the real axis and to exclude those on the negative part, these real poles contribute to unattenuated outgoing waves. Therefore, it will suffice to determine those roots of (C.17) that have either a strictly positive imaginary part or, in the case of real roots, are strictly positive. It has been shown, cf. [8], that (C.17) has only one positive real root, corresponding to an unattenuated, outgoing wave, accompagnied by a surface wave in the fluid, travelling to infinity at subsonic speed. Moreover, it is easy to show that real roots of (C.17) must be greater than $\max(1, M)$, which means that the surface wave travels at *subflexural* rather than *subsonic* speed for frequencies below

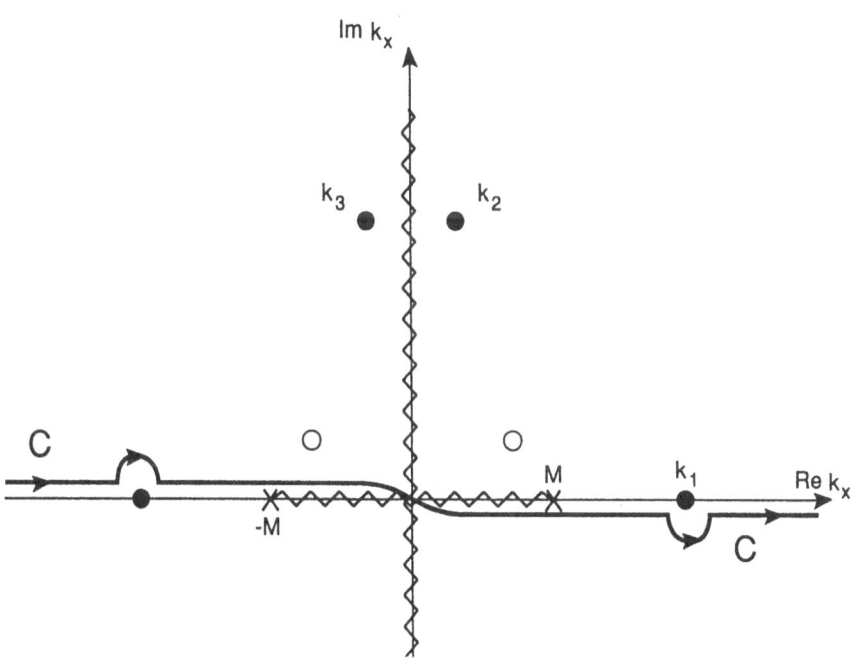

FIGURE C1. Contour of integration C for evaluating inverse Fourier transforms. $\times$ branch points of $\bar{k}_z(\bar{k}_x)$; $\bullet$ poles of $1/(Z_p + Z_a)$; $\circ$ poles of $1/(Z_p - Z_a)$; $\sim\!\!\sim\!\!\sim$ branch cuts of $\bar{k}_z(\bar{k}_x)$.

coincidence. It is noted here that the frequency range where the classical plate theory applies is restricted to frequencies below coincidence. Moreover, this criterion happens to be a practical tool for excluding spurious roots if one uses a library routine for solving the polynomial equation (C.16). The number and the location of the other roots of (C.17) depend on the choice of the branch cut for the function $\bar{k}_z = (M^2 - \bar{k}_x^2)^{\frac{1}{2}}$. This choice is not unique. A convenient choice is to take an L-shaped branch cut, running along the real axis from M to 0 and continuated along the imaginary axis from $i0$ to $+i\infty$. For this choice the square root function $\bar{k}_z$ is either purely real or purely imaginary, so complicated formulas in the branch cut integral are avoided. Straightforward analysis of (C.17) brings up a useful criterion for discriminating spurious complex-valued roots of (C.16) from those of (C.17):

$$\text{Im}(\bar{k}_x) > (1 + \sqrt{2})|\text{Re}(\bar{k}_x)|.$$

The contour C, the branch cuts of $\bar{k}_z$ and the poles of (C.17) that contribute to residue contributions are sketched in Figure C1. The real pole is denoted by $\bar{k}_1$, the complex ones by $\bar{k}_2$ and $\bar{k}_3$. It is noted that $\bar{k}_3 = -\bar{k}_2^*$.

The final expression for the plate response takes the form, cf. [7,9] :

$$\bar{u}(\bar{x}) = \frac{u(x)}{(-i\omega)^{-1}k_p F_0'/4m\omega} = 4\,[R_{1u} + R_{2u} + R_{3u} + B_{1u} + B_{2u}], \quad \text{(C.18)}$$

where the residues R_{1u}, R_{2u} and R_{3u} and the two parts of the branch cut integral B_{1u} and B_{2u} are given by

$$R_{1u} = \frac{\exp(i\bar{k}_1\bar{x})}{4\bar{k}_1^3 + (\varepsilon/M)\bar{k}_1(\bar{k}_1^2 - M^2)^{-3/2}}, \quad \text{(C.19a)}$$

$$R_{2u,3u} = \frac{\exp(i\bar{k}_{2,3}\bar{x})}{4\bar{k}_{2,3}^3 - i(\varepsilon/M)\bar{k}_{2,3}(M^2 - \bar{k}_{2,3}^2)^{-3/2}}, \quad \text{(C.19}b,c\text{)}$$

$$B_{1u} = -\frac{i}{\pi}\frac{\varepsilon}{M}\int_0^\infty dq \frac{\sqrt{M^2 + q^2}\exp(-q\bar{x})}{(q^4 - 1)^2(M^2 + q^2) + \varepsilon^2/M^2}, \quad \text{(C.19}d\text{)}$$

$$B_{2u} = \frac{1}{\pi}\frac{\varepsilon}{M}\int_0^M dp \frac{\sqrt{M^2 - p^2}\exp(ip\bar{x})}{(p^4 - 1)^2(M^2 - p^2) + \varepsilon^2/M^2}. \quad \text{(C.19}e\text{)}$$

It is noted that $R_{3u} = R_{2u}^*$, so $R_{2u} + R_{3u} = 2\,\text{Re}[R_{2u}]$. Furthermore, the integral for B_{1u} is of the Laplace-transform type of a non-oscillating function, which is easily evaluated numerically. It is noted that the derivatives of $\bar{u}(\bar{x})$, i.e. $\bar{u}'(\bar{x})$ and $\bar{u}''(\bar{x})$ can be obtained directly from differentiating these expressions with respect to $\bar{x}$.

Finally, it is noted once again that the nondimensional distance $\bar{x}$ in (C.18) and (C.19) differs from $\bar{x}$ in the text, since it is scaled by k_p^{-1} instead of a. Similarly, $\bar{u}$ from (C.18) and $\bar{\gamma}$ from the text have different scalings too.

D Some properties of the Fourier integral transform

Consider the Fourier transform pair

$$\hat{f}(\kappa) = \int_{-\infty}^{\infty} \bar{f}(\bar{x}) \exp(-i\kappa\bar{x})d\bar{x}, \tag{D.1}$$

$$\bar{f}(\bar{x}) = \frac{1}{2\pi} \int_{-\infty}^{\infty} \hat{f}(\kappa) \exp(i\kappa\bar{x})d\kappa. \tag{D.2}$$

The following relations hold for the Fourier transforms of functions with a shifted argument and for derivatives of such functions:

$$\int_{-\infty}^{\infty} \bar{f}(p\bar{x} + q) \exp(-i\kappa\bar{x})d\bar{x} = \frac{1}{|p|} \exp[i(q/p)\kappa]\hat{f}(\kappa/p), \tag{D.3}$$

$$\int_{-\infty}^{\infty} \bar{f}'(p\bar{x} + q) \exp(-i\kappa\bar{x})d\bar{x} = \frac{i\kappa}{p|p|} \exp[i(q/p)\kappa]\hat{f}(\kappa/p). \tag{D.4}$$

For the convolution $\bar{c}(\bar{x})$ of two functions, we have

$$\bar{c}(\bar{x}) = \int_{-\infty}^{\infty} \bar{f}(\bar{x} - \bar{\xi})g(\bar{\xi})d\bar{\xi} \equiv \int_{-\infty}^{\infty} \bar{f}(\bar{\xi})g(\bar{x} - \bar{\xi})d\bar{\xi}, \tag{D.5}$$

$$\hat{c}(\kappa) = \int_{-\infty}^{\infty} \bar{c}(\bar{x}) \exp(-i\kappa\bar{x})d\bar{x} = \hat{f}(\kappa)\hat{g}(\kappa). \tag{D.6}$$

Mathematical modeling and dimensional analysis

An application to four-wheel steering

J. Molenaar
Institute for Mathematics Consulting
Faculty of Mathematics and Computing Science
Eindhoven University of Technology
P.O. Box 513, 5600 MB Eindhoven, The Netherlands

1. Introduction

We present in §2 a general introduction on mathematical modeling. Irrespective of the case under consideration, the process of mathematical modeling has a general structure. See also references [1,–,6]. This structure comprises a sequence of stages followed again and again. An important ingredient of the modeling process is *dimensional analysis*. In §3 the main ideas of this technique are presented.

To illustrate the general concepts in §§2 and 3 we deal in §4 with the modeling of four-wheel steering. It will turn out, that we have to face the general question which path is followed by the back-wheels of a vehicle when the front-wheels are steered along a given path. The model to be developed is not only of importance in car industry. Its answer can be used to determine, e.g., the position of vehicles while turning and this knowledge is necessary in designing road plans, parking-places, etc. Furthermore, in most countries long vehicles have to meet certain requirements with respect to their kinematical behaviour. Instead of meeting these requirements by trial and error, it is often cheaper and quicker to apply mathematical modeling.

A. van der Burgh and J. Simonis (eds.), Topics in Engineering Mathematics, 93–119.

2. Mathematical Modeling

If real-life problems are attacked using mathematics, a "translation" is needed to put the subject into a mathematically tractable form. A possible definition of this process, usually called mathematical modeling, reads: *mathematical modeling is the description of an experimentally verifiable phenomenon by means of the mathematical language.* The phenomenon to be described will be called *the system*, and the mathematics used, together with its interpretation in the context of the system, will be called *the mathematical model.* In most mathematical models we find two classes of quantities:

– **Variables**
  Within this class we distinguish dependent from independent variables. E.g., one may be interested in the temperature (dependent variable) of a system as a function of time and position (independent variables). Or, one may look for the position (dependent variable) of an object as a function of time (independent variable). If the dependent variables are differentiable functions of the independent ones, the model might comprise a set of differential equations. The number of independent variables then determines whether one has to do with ordinary differential equations (one independent variable) or with partial differential equations (more than one independent variables).

– **Parameters**
  In this class we distinguish parameters, which are practically constant, from parameters, which can be adjusted by the experimenter. The acceleration of gravity and the decay time of radio-active materials are examples of the first kind. The temperature of chemical reactions is an example of the second kind: the experimenter can often control the reaction by means of the environmental temperature, because in many systems the reaction rates strongly depend on this parameter.

A mathematical model is said to be *solved* if the dependent variables are in principle known as functions of the independent variables and the parameters. The *solution* may be obtained either analytically or numerically.

The process of mathematical modeling generally consists of the following steps:

1: *Orientation*
2: *Formulation of relations between variables and parameters*
3: *Non-dimensionalization*
4: *Reduction*
5: *Mathematical analysis*
6: *Verification*
7: *Reiteration of steps 2–6*
8: *Implementation*

We discuss these stages separately:

Step 1: The modeling process always starts with an orientation stage, in which the modeler gets acquainted with the system under consideration by means of observations and/or information from experts and the literature. Large scale experiments are not yet relevant at this stage. In fact, most experiments are premature if they are not based on some model. E.g., a common mistake in the field of statistics, is that much energy is spent on gathering data before any model is developed. Testing of models developed afterwards frequently leads to the conclusion that some essential parameter is unfortunately not measured.

Step 2: The next step is to formulate relations between variables and parameters. For certain classes of systems these relations have been established already long ago. We mention, e.g., the laws of Newton in classical mechanics, the Maxwell equations in electromagnetism, and the Navier-Stokes equations in fluid mechanics. In many other cases rules of thumb are in use, which do not have the same status as the well-accepted fundamental laws. These rules may certainly be reliable and useful, but always only in a restricted context only. It may also happen that the modeler has to start from scratch. An important condition is that of *consistency*: the proposed relations may not contain contradictions. E.g., in case of differential equations one has to specify not too few and not too many boundary and/or initial conditions in order to ensure that the model has a unique solution.

Step 3: This step is extensively discussed in Appendix A. See also, e.g., references [7,8].

Step 4: Most models are in the first instance not tractable. It therefore makes nearly always sense to look for reduced models, that are still reliable but only under certain restrictions. Such simplified models are often very useful. E.g., a simple model can already make clear that it is possible to bring a rocket into an orbit around the moon, whereas the complexity of the calculations needed to evaluate the full model is extremely high. The technique of reducing a model requires much experience, because one has to neglect many details meanwhile retaining the essential aspects. E.g., friction is often neglected in mechanical models, but in a model for the clutch or the brakes of a car it might be the heart of the matter.

Step 5: In the mathematical analysis of a model one tries to find the solution by means of one of the methods from the huge reservoir of mathematical knowledge developed in history. If no existing method applies, one either has to develop new mathematical tools or one has to go back to step 4.

Step 6: If an analytical or numerical way has been found to solve the (reduced) model, it remains to explore the solution as a function of the independent variables and the adjustable parameters. Not all possibilities are of practical use. In most cases only certain tendencies or special points are relevant. These features have to be checked against data. At this stage the modeler may propose to perform specific experiments. A model that is too complex to be verified experimentally is useless. Also, a model that contains so many parameters that nearly all kinds of data can be fitted is not interesting from a practical point of view.

Step 7: After having compared the measured data and the calculated solution, one often has gained enough insight to improve the model and to reiterate steps 2–6. In view of this iteration process it is convenient to follow a *modular* approach from the very beginning; i.e., one rather should start with the most simple, but still relevant model, gradually adding more and more aspects.
It scarcely happens that a model describes all dependent variables equally well. Of course, the specification of the "value" of a model is a subtle matter, which may vary from modeler to modeler. In practice, the user and not the designer of a model will determine when the iteration process may be stopped.

Step 8: If a model suffices for a specific application, the implementation phase starts. In general the results are used by non-mathematicians, who need instruction about the power and the poverty of the model. The appropriate presentation of the results is an important part of each project, and the attention and time it requires should not be underestimated.

We close this section with some general remarks:

- One and the same system may be described satisfactorily by different mathematical models. These models may complement each other, as is the case e.g., with the particle and the wave models for light and elementary particles. Identification of a system with one particular model is a serious misunderstanding of the character of "modeling". Another misinterpretation, often met in popular scientific publications, is the statement that a phenomenon is "understood" if a satisfactory model exists. Mathematical models are concerned about "how" things happen, and not "why" things occur.

- Not all systems can be suitably modeled by means of mathematics. Outspoken examples of systems appropriate for this approach are those studied in physics and chemistry. This is less and less the case with the systems studied in, e.g., biology, economy, social sciences, and politicology, respectively. Many, if not most, real-life systems are governed by laws, which can hardly be put into mathematical form.

- Mathematical modeling is, in the first instance, merely descriptive and it therefore comes under fundamental research. However, all successful models will, sooner or later, be used in technical applications for the design and control of similar systems.

- One and the same model may be useful with respect to different systems. This may become apparent after non-dimensionalization.

3. Dimensional Analysis

The variables and parameters in a mathematical model have in general physical dimensions. The dimensions of most of them will be obvious (time, length, mass, temperature, etc.), and the dimensions of the remaining ones can be deduced from the rule that all terms in one equation must have the same dimensions. This rule stems from the condition that no equation may depend on the units used to measure the dimensions. From these considerations the dimensionality of constants of proportionality follow directly. E.g., if a friction force is introduced with its strength linearly proportional to the velocity of the object under consideration, then the constant of proportionality will have the dimensions of the quotient of force and velocity.

Concise introductions into dimensional analysis are references [7,8]. Non-dimensionalization of a model first implies that a (non-linear) transformation is applied to the set of variables and parameters, which yields dimensionless combinations of them. Next, the model is made dimensionless by rewriting all equations in terms of these dimensionless quantities. It is by no means clear in advance that this always possible, but the existence of such a transformation is proved by the theorem given below. In this theorem it is at the same time pointed out how such transformations can be found. The technique of non-dimensionalization is an extremely powerful tool in mathematical modeling. Its importance is appreciated best when one applies the method. We therefore present several examples at the end of this appendix. First, we wish to summarize some striking advantages:

- The number of variables and parameters decreases.

- Dimensional analysis may yield insight in the general scaling properties of the system. In Example 1 underneath this point is illustrated by deriving the famous theorem of Pythagoras by means of arguments from dimensional analysis.

- Mathematical models that describe completely different systems look like quite different in the first instance. However, sometimes such models appear to be identical if being put into dimensionless form.

- The reduction of a model (Step 4 in the modeling scheme of §2) is often accomplished by neglecting those terms in the model equations which are much smaller than the remaining terms. It should be realized, however, that comparison of the magnitudes of terms only makes sense if the model is in dimensionless form.

- Quite often it is attractive to perform experiments on systems that have been scaled down in size. Only non-dimensionalization of the model can make clear whether the results of such experiments are still meaningful with respect to the original system. We illuminate this point in Example 3 underneath.

Now we turn to the central theorem of dimensional analysis. In this theorem we only deal with scalar variables and parameters. The components of vector-valued variables and parameters are thus treated separately.

Theorem (Buckingham). *We consider a system with (scalar) variables* $x_1, \ldots, x_k$, *and (scalar) parameters* $p_1, \ldots, p_l$. *The associated dimensions are denoted by* $d_1, \ldots, d_m$. *Each relation*

$$f(x_1, \ldots, x_k, p_1, \ldots, p_l) = 0$$

can be rewritten in the equivalent, dimensionless form

$$\bar{f}(q_1, \ldots, q_n) = 0$$

with $q_1, \ldots, q_n$ *dimensionless products of (powers of) the* x's *and* p's. *The number* n *is given by*

$$n = k + l - m .$$

Proof. We write $M = k + l$. Let us introduce the set V with elements v of the form

(a.1) $\quad v = x_1^{r_1} \ldots x_k^{r_k} \, p_1^{r_{k+1}} \ldots p_l^{r_M}$

with $r_i \in \mathbb{R}$, $i = 1, \ldots, M$. There is an obvious one-to-one correspondence between the elements of V and the vectors $(r_1, \ldots, r_M) \in \mathbb{R}^M$. The corresponding mapping is denoted by $T_1 : \mathbb{R}^M \to V$. If the x's and p's in the right hand side of (a.1) are replaced by their associated dimensions, a set W is obtained with elements w of the form

$$w = d_1^{s_1} \ldots d_m^{s_m}$$

with $s_i \in I\!R$, $i = 1, \ldots, m$. This replacement procedure induces a mapping $T_2 : V \to W$. This mapping is surjective, because each dimension occurs in the system. There is an obvious one-to-one correspondence between the elements of W and the vectors $(s_1, \ldots, s_m) \in I\!R^m$. This mapping is denoted by $T_3 : W \to I\!R^m$. The composite mapping

$$T = T_3 T_2 T_1 \quad ; \quad T : I\!R^M \to I\!R^m$$

is linear and surjective. Its null space $N_0 \subset I\!R^M$ has dimension $n = M - m = k + l - m$. The elements of N_0 just correspond to the dimensionless elements of W. We choose a basis $q_1, \ldots, q_n$ in N_0 and extend it to a basis in $I\!R^M$ by adding linearly independent elements $q_{n+1}, \ldots, q_M$. All x's and p's can be written as unique, linear combinations of these basis elements. This implies that every relation $f(x_1, \ldots, x_k, p_1, \ldots, p_l) = 0$ can uniquely be rewritten in the form $\bar{f}(q_1, \ldots, q_M) = 0$. However, the function $\bar{f}$ has to be independent of the units used to measure the dimensions. This can only be the case if $\bar{f}$ does not contain any of the basis elements $q_{n+1}, \ldots, q_M$, because their values may attain any value if the units are varied. Thus $\bar{f} = \bar{f}(q_1, \ldots, q_n)$, and this completes the proof. $\square$

The following points should be realized if this theorem is applied in practice:

- The choice of the q's is not unique in most systems. Different choices may lead to quite different dimensionless models and the mathematical analysis of one model (Step 5 in §2) may be much more convenient than that of another.

- A relatively simple version of non-dimensionalization is often already quite effective. In this approach a transformation is applied which makes each variable dimensionless by dividing it by a convenient combination of parameters. E.g., all variables with the dimension of length are divided by a characteristic length of the system, all variables with the dimension of time are divided by a characteristic time of the system, etc. Non-dimensionalization is then nothing else but a scaling of the variables by means of the parameters, and the number of q's is equal to the number of the variables. We illustrate this in Example 4.

– Dimensional analysis yields more insight if it is possible to find a set of q's such that at least one q contains more than one variable. The corresponding transformation is sometimes called a *similarity transformation*. See also references [9,10]. In Example 4 such a transformation is explicitly shown.

Example 1.
From [7] we take the following, remarkable example. The theorem of Pythagoras can be derived by means of arguments from dimensional analysis. We consider a right-angled triangle ABC as drawn in the figure.

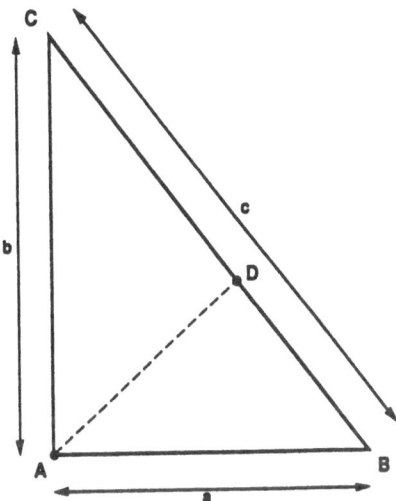

Such a triangle can be fixed by specifying two of its sides, say a and c. This system has no variables and two parameters, both with the dimension of length. So, there is only one dimensionless quantity, and an obvious choice is

$$q \equiv a/c .$$

The area O of the triangle is a function of a and c and we may write

$$O = f(a,c) = c^2(f(a,c)/c^2) .$$

Because both O and c^2 have the dimensions of length squared, we conclude that the quotient f/c^2 is dimensionless and can only depend on q. So, we may introduce the notation

$$f/c^2 \equiv \bar{f}(q) \quad ; \quad O = c^2\,\bar{f}(q)\,.$$

This reasoning holds for all triangles congruent with triangle ABC. All these triangles have the function $\bar{f}$ and the value of q in common. So, for these triangles is the factor $\bar{f}(q)$ a constant, say $\bar{f}_0$. As seen in the figure, triangles DBA and DAC are congruent with the original triangle ABC. We may thus write

$$O_{ABC} = c^2\,\bar{f}_0 = O_{DBA} + O_{DAC} = a^2\,\bar{f}_0 + b^2\,\bar{f}_0\,,$$

from which the famous relation $a^2 + b^2 = c^2$ follows directly.   □

Example 2.
Here, we study the consequences of dimensional analysis if applied to a mathematical pendulum. We restrict the movement of the pendulum to a vertical plane. In practice the swinging behaviour of the pendulum will damp out because of friction. We assume the strength of the friction force to be directly proportional to the velocity squared.

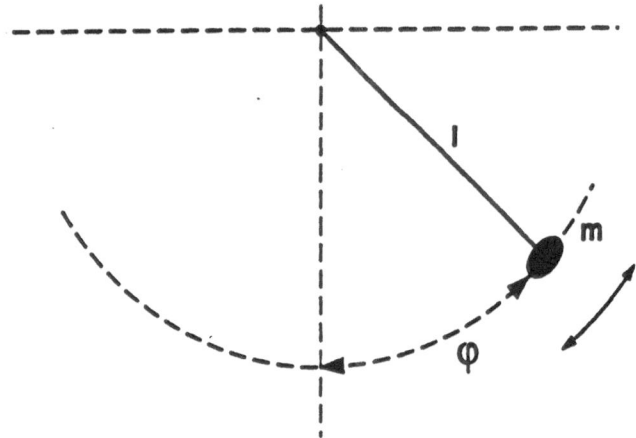

What are the variables, parameters, and dimensions of this system?
The independent variable is the time t. Its dimension is denoted by $[t] \equiv T$. The dependent variable is the dimensionless angle φ, indicated in the Figure. The parameters are the length l of the pendulum with $[l] \equiv L$, its mass m with $[m] \equiv M$, the acceleration of gravity g with $[g] = L/T^2$, the dimensionless, initial angle φ_0 ($\neq 0$), and the friction coefficient α. The

dimensions of α are given by the quotient of force and velocity squared, thus $[\alpha] = M/L$. In the system, modeled this way, 7 variables and parameters and 3 dimensions are involved, so the number of dimensionless quantities q will be 4. Because φ and φ_0 are already dimensionless, obvious choices are $q_1 \equiv \varphi$, and $q_2 \equiv \varphi_0$. To find q_3 and q_4 we form the products

$$q = t^{r_1}\, l^{r_2}\, m^{r_3}\, \alpha^{r_4}\, g^{r_5} \ .$$

The condition $[q] = 0$ leads to three linear equations for $r_1, \ldots, r_5$:

$$r_1 - 2r_5 = 0$$

$$r_2 - r_4 + r_5 = 0$$

$$r_3 + r_4 = 0 \ .$$

We may choose two r's freely. The natural choices $(r_1, r_2) = (1, 0)$ and $(r_1, r_2) = (0, 1)$ yield $q_3 = t\,\sqrt{\alpha g/m}$ and $q_4 = l\alpha/m$ respectively. Each property of this system can thus be expressed by an equation of the general form

$$(a.2) \qquad f(q_1, \ldots, q_4) = f(\varphi_1, \varphi_0, t\,\sqrt{\alpha g/m},\ l\alpha/m) = 0 \ .$$

Our first conclusion is, that all pendulums of given length and given ratio α/m behave identically, if started at the same φ_0.

Let us next rewrite relation (a.2) in the form

$$q_3 = \bar{f}(q_1, q_2, q_4) \ ,$$

i.e.,

$$(a.3) \qquad t = \sqrt{\frac{m}{\alpha g}}\ \bar{f}(\varphi_1, \varphi_0, l\alpha/m) \ .$$

We note, that this inversion will not be possible for all t. From the physics of the system we know, that φ is a univalued function of t, but this will hold for t as a function of φ only as long as the pendulum has not yet changed its direction of motion.

Let us direct our attention to the time $t_{1/2}$, at which the amplitude of the pendulum has been halved for the first time. Thus, $\varphi(t_{1/2}) = \frac{1}{2}\varphi_0$. From (a.3) we may write

$$t_{1/2} = \sqrt{m/\alpha g}\ \bar{f}(\varphi_0, l\alpha/m)\ .$$

Our second conclusion from dimensional analysis is that, if we vary the values of $l\alpha$ and m, but keep the value of the quotient $l\alpha/m$ fixed, $t_{1/2}$ scales with $\sqrt{m/\alpha}$ or, equivalently, with $\sqrt{l}$.

Let us reduce the model by neglecting the friction. This implies that $r_4 = 0$, and thus $r_3 = 0$. From this a third conclusion follows: the behaviour of a frictionless pendulum is independent of its mass. For the reduced model we find unambiguously $q_3 = t\sqrt{g/l}$, and we may write

$$t = \sqrt{l/g}\ \bar{f}(\varphi, \varphi_0)\ .$$

Dimensional analysis itself does not tell us that, in the frictionless system, φ is a periodical function of time. If we take this for granted, our fourth conclusion is that the period τ satisfies

$$\tau = \sqrt{l/g}\ \bar{f}(\varphi_0)$$

and thus scales with $\sqrt{l/g}$. It requires the explicit solution of the equation of motion to find that $\bar{f}(\varphi_0)$ is given by an elliptic integral, which reduces to 2π if $|\varphi_0| \ll 1$.

Example 3.

We wish to model a ship sailing at constant speed. Obvious parameters are its length l with dimension $[l] \equiv L$ and its velocity v with $[v] = L/T$. The movement of the ships sets the water in motion because water is viscous. The viscosity α has dimensions $[\alpha] = M/LT$. As a result of the viscous friction, energy is transferred from the ship to the water. This energy is used partly to induce surface waves and partly to generate turbulent motion of the water. Because of these effects also the acceleration of gravity g, with $[g] = L/T^2$, and the density of water ρ, with $[\rho] = M/L^3$, will play a role.

If we assume that the ship is streamlined such that its height and width are not of importance in the present analysis, the system has five variables and parameters. Because three dimensions are involved, the number of q's is two. As above, we form the products

$$q = v^{r_1} \rho^{r_2} l^{r_3} \alpha^{r_4} g^{r_5} .$$

The condition $[q] = 0$ yields three equations:

$$r_1 - 3r_2 + r_3 + r_5 = 0$$

$$r_1 + r_4 + 2r_5 = 0$$

$$r_2 + r_4 = 0 .$$

The choice $(r_1, r_2) = (1, 0)$ yields $q_1 = v/\sqrt{lg}$. This is called the *Froude number* after William Froude, a famous ship builder. The choice $(r_1, r_2) = (0, 1)$ yields $\bar{q}_2 = \rho l \sqrt{lg}/\alpha$. For historical reasons one prefers to introduce $q_2 = q_1 \bar{q}_2 = v\rho l/\alpha$. The latter dimensionless quantity is called the *Reynolds number*, after Osborne Reynolds, a researcher in fluid mechanics. Because real-life experiments are hard for these systems, it is very attractive to perform experiments on (physical) models in which all sizes are scaled down by a certain factor. The conclusions from these experiments are valid for the original system only if both systems are described by the same dimensionless (mathematical) model. So, q_1 and q_2 have to remain constant upon scaling. In practice the values of g, ρ, and α can hardly be adjusted. To keep q_1 constant, $v/\sqrt{l}$ may not change, and to keep q_2 constant, vl must be preserved. These requirements cannot be satisfied at the same time. So, in the first instance the conclusion is that (physical) scaling has no sense. However, under certain conditions scaling may be still useful. If the generation of surface waves is unimportant compared to the other mechanisms of energy dissipation, we may reliably ignore the Froude number. In that case one only has to keep the Reynolds number constant, which implies that the velocity of the scaled down model must be larger than the velocity of the real system. On the contrary, if the Froude number is much larger than the Reynolds number, the latter may be ignored. Then, the speed of the scaled down model must be smaller than that of the original ship.

Example 4.

We consider heat diffusion in a long rod. The rod is thermally isolated everywhere except one of the end points. The system acts as a one-dimensional conductor. Initially the rod is uniformly at temperature τ_0. From a certain moment on the end of the rod is brought into contact with a heat reservoir, which keeps that end of the rod at constant temperature τ_1. We are interested in the temperature $\tau(t, x)$ in the rod as a function of time t and position x. This well-known system is described by the heat diffusion equation

$$(\text{a.4}) \qquad \frac{\partial \tau}{\partial t} = k \frac{\partial^2 \tau}{\partial x^2}$$

with boundary condition

$$\tau(t, 0) = \tau_1 , \quad t \geq 0$$

and initial condition

$$\tau(0, x) = \tau_0 , \quad 0 < x \leq l .$$

The independent variables are time t with $[t] \equiv T$, and position x with $[x] \equiv L$. The dependent variable is the temperature τ with $[\tau] \equiv \text{TMP}$. The parameters are the length of the rod l with $[l] \equiv L$, the temperatures τ_0 and τ_1 with $[\tau_0] = [\tau_1] = \text{TMP}$, and the thermal conductivity k with $[k] = L^2/T$. An obvious way to obtain dimensionless quantities is to scale the variables using characteristic quantities for length, time, and temperature, respectively:

$$q_1 = x/l$$

$$q_2 = t/(l^2/k)$$

$$q_3 = (\tau - (\tau_1 - \tau_0)) / (\tau_1 - \tau_0) .$$

In the mathematical model it may make sense to assume the rod to be infinitely long. In practice, this assumption is reasonable if l^2/k is much larger than the period during which we wish to observe the system. For $l \to \infty$ dimensional analysis gives rise to a far reaching conclusion. As above, we form the products

$$q = t^{r_1} \, x^{r_2} \, \tau^{r_3} \, k^{r_4} (\tau_1 - \tau_0)^{r_5} \ .$$

Because the system is linear, only the difference $\tau_1 - \tau_0$ will play a role, and not τ_1 and τ_0 separately. From the condition $[q] = 0$ we obtain

$$r_1 - r_4 = 0$$

$$r_2 + 2r_4 = 0$$

$$r_3 + r_5 = 0 \ .$$

From these equations it follows that

$$q_1 = x^2/kt$$

$$q_2 = (\tau - (\tau_1 - \tau_0)) / (\tau_1 - \tau_0) \ .$$

Because the physics of the system is such, that the temperature $\tau = \tau(t, x)$ will be a unique function of the pair (t, x), we may write

(a.5) $q_2 = f(q_1)$,

or

$$\tau = (\tau_1 - \tau_0) \left(f(x^2/kt) + 1 \right) \ .$$

The important conclusion is that the solution $\tau(t, x)$ of (a.4) only depends on the quotient x^2/kt if the rod may be taken infinitely long. This implies that the partial differential equation (a.4) passes into an ordinary differential equation if written in dimensionless form. It appears that the function $f(q)$ in (a.5) satisfies the equation

$$4q \, \frac{\partial^2 f}{\partial q^2} + (q + 2) \, \frac{\partial f}{\partial q} = 0 , \quad 0 < q < \infty$$

with the boundary conditions

$$f(0) = 0 , \quad f(\infty) = 1 \ .$$

The solution is the complementary error function

$$f(q) = \text{erfc}\left(\tfrac{1}{2}\sqrt{q}\right) = \frac{1}{\sqrt{\pi}} \int\limits_{\frac{1}{2}\sqrt{q}}^{\infty} e^{-s^2}\, ds\;.$$

4. Four-Wheel Steering

Why is it difficult to drive a long vehicle backward into a gate? Why is parking a hard job if the parking space is limited? Is rear-wheel steering favourable when driving around a corner? These questions have one central theme in common: what is the relation between the rear-wheel and the front-wheel trajectories of a vehicle? To answer this question, we shall work out a model system. We start here modeling steps 1: *Orientation*, and 2: *Formulation of relations between variables and parameters.*

In case of long vehicles it seems reasonable to assume that the number of front and rear wheels is of only minor importance, and hence we restrict our model to have only one front and one rear wheel. Both are free to turn. We assume that the friction between wheel and road-surface is so large that slip angles may be neglected. The model vehicle is sketched in Fig. 1. The front and back wheels touch the ground at the points A and B respectively, and the distance between these points is l. To simulate rear-wheel steering, we assume that the rear wheel turns through an angle $\alpha\varphi$, where $-1 < \alpha < +1$, if the front wheel turns through an angle φ to the right or the left. For $\alpha > 0$ front and rear wheels both steer to the same side, whereas for $\alpha < 0$ they turn in opposite directions. For $\alpha = 0$ the vehicle resembles a bicycle.

Given the trajectory of the front wheel, the derivation of the equation of motion of the rear wheel requires the application of some differential geometry in the plane. General introductions to differential geometry are references [11,12,13]. The equation of motion will be obtained from the observation that the velocity of the rear wheel must always be tangential to the trajectory of the rear wheel.

We denote the path of the front wheel by $\mathbf{x}(s)$, where s is the arc length. In $\mathbf{x}(s)$ we choose a local, orthonormal coordinate system $(\mathbf{t}, \mathbf{n})$, where $\mathbf{t}$ is the tangent and $\mathbf{n}$ the normal vector. The reader's attention is drawn to the difference between time t and tangent $\mathbf{t}$. To fix the orientation we take the direction of $\mathbf{n}$ pointing to the left hand side of the driver. This implies that the front-wheel path has positive curvature if the driver turns to the left.

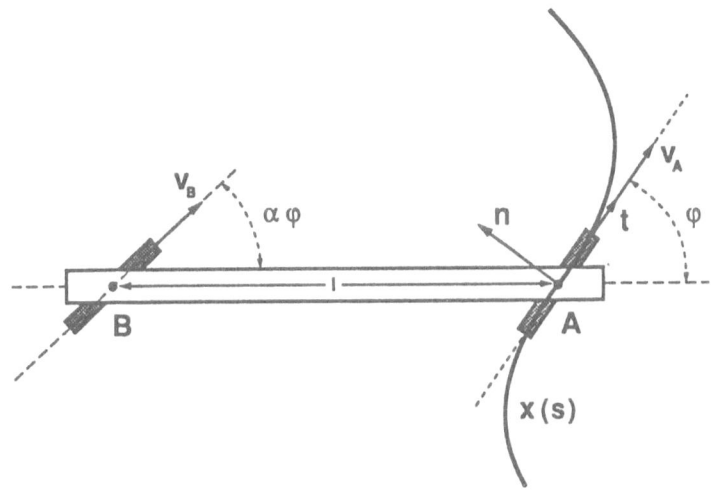

Fig. 1.

The tangent **t** and normal **n** are given by

$$t = \frac{d}{ds} \mathbf{x} , \quad \|t\| = 1$$

$$n = \frac{1}{k} \frac{d}{ds} t , \quad \|n\| = 1 .$$

with k the (s-dependent) curvature of the path $\mathbf{x}(s)$.
The velocity $\mathbf{v}_A$ (with absolute value v_A) of the front wheel satisfies the equation

$$\mathbf{v}_A = \frac{d\mathbf{x}_A}{dt} \equiv \dot{\mathbf{x}}_A = \dot{s} \frac{d\mathbf{x}_A}{ds} = v_A t ,$$

where $ds/dt \equiv \dot{s} \equiv v_A$. In the theory that follows we shall need the time derivatives $\dot{t}$ and $\dot{n}$. Because t and n are unit vectors, they satisfy the (2-dimensional) Serret-Frenet formulae $dt/ds = kn$ and $dn/ds = -kt$. We then obtain the expressions

(3.1)
$$\begin{cases} \dot{t} = \dot{s} \frac{dt}{ds} = v_A k\, n \\[2mm] \dot{n} = \dot{s} \frac{dn}{ds} = -v_A k\, t . \end{cases}$$

The position $\mathbf{x}_B$ of the rear wheel is given by

(3.2a) $\mathbf{x}_B = -l(\cos \varphi \mathbf{t} + \sin \varphi \mathbf{n})$,

in which φ is taken to be positive in the first quadrant of $(\mathbf{t}, \mathbf{n})$. By differentiating with respect to time t we find an equation for the velocity $\mathbf{v}_B$:

(3.2b) $\mathbf{v}_B = (v_A + l \sin \varphi(\dot{\varphi} + k\, v_A))\mathbf{t} - l \cos \varphi(\dot{\varphi} + k\, v_A)\mathbf{n}$.

The equation of motion is obtained from the condition that this velocity is tangential to the path of the rear wheel, i.e., $\mathbf{v}_B$ should be parallel to $\mathbf{t}_B$. From Fig. 2 we see that $\mathbf{t}_B$ is given by

(3.2c) $\mathbf{t}_B = \cos(1 - \alpha)\varphi \mathbf{t} + \sin(1 - \alpha)\varphi \mathbf{n}$.

Equating the first and second components of the right-hand-sides of (3.2b) and (3.2c) we arrive at the autonomous equation of motion

(3.3) $\dot{\varphi} = -\dfrac{v_A}{l}\left(\sin \varphi - \tan \alpha\varphi \cos \varphi + lk\right)$.

Let us turn now to modeling step 3: *Non-dimensionalization.*
The dimensional quantities in this equation are time t, length l, velocity v_A and curvature k. By means of the dimensional analysis presented in Appendix A we may derive the following dimensionless parameters:

$$\begin{cases} t^* = v_A\, t/l \\ k^* = lk\ . \end{cases}$$

Note, that $t^* = s/l$ if v_A is constant. In that case t^* stands for arc length measured in units of length l. In the following we shall restrict ourselves to this case for convenience. This implies that the model becomes independent of v_A.
Omitting the $*$ superscript the dimensionless equation of motion reads

(3.4) $\begin{cases} \dot{\varphi} = -(\sin \varphi - \tan \alpha\varphi \cos \varphi + k) \\ \varphi(0) = \varphi_0\ . \end{cases}$

To get a feeling for the behaviour of its solution we shall analyse (3.4) for some special cases. So, we now enter modeling steps 4: *Reduction* and 5: *Mathematical analysis.*

Straight line

We first consider the case that the front wheel is driven along a straight line, so $k = 0$. At $t = 0$ the vehicle forms an angle φ_0 with this line.

For $\alpha = 0$ (3.4) reduces to

$$\dot{\varphi} = -\sin \varphi$$

with stationary points $\varphi = 0, \pi$. From the Jacobian $J(\varphi) = -\cos \varphi$ we find directly that $\varphi = 0$ is a stable point and $\varphi = \pi$ an unstable point. Driving backwards is an unstable activity! The solution $\varphi(t)$ is readily obtained by separation of variables:

$$\int_{\varphi_0}^{\varphi(t)} \frac{d\psi}{\sin \psi} = -\int_0^t d\tau = -t .$$

Thus (see, e.g., reference [15])

$$\varphi(t) = 2\arctan(e^{-t} \tan(\tfrac{1}{2} \varphi_0)) .$$

For $\alpha \neq 0$ we have

(3.5) $\dot{\varphi} = -\sin \varphi + \tan \alpha\varphi \, \cos \varphi .$

The stationary points satisfy the equations

$$\tan \varphi = \tan \alpha\varphi .$$

or

$$\sin(1 - \alpha)\varphi = 0 \ .$$

If $\alpha = 1$ each φ is stationary, but for $\alpha \neq 0$ the situation is quite complex. In that case the points $\varphi_i = i\pi/(1 - \alpha)\bmod(2\pi)$, $i = 1, 2, \ldots$ are stationary. If α is rational, the number of φ_i is finite, but for α irrational this number is infinite.

These results make it clear that we must *interpret* our model anew. It appears that the model – as it stands – admits quite peculiar solutions due to the fact that the front wheel may turn around without limit. We therefore add the restriction

$$|\varphi| \leq \varphi_m \ .$$

with $\varphi_m < \pi$ a given maximum angle. Under this restriction the only stationary point is $\varphi = 0$. The situation now is that for some initial conditions φ_0 (with $|\varphi_0| < \varphi_m$) the solution $\varphi(t)$ will tend to zero for $t \to \infty$, while for other values of φ_0 the solution $\varphi(t)$ will reach the upper bound φ_m and the vehicle will stop abruptly.

Circle

In this case the path of the front wheel is a circle with (constant) curvature k. If the radius of the circle is denoted by R, we have $k = l/R$.

For $\alpha = 0$ (3.4) reduces to

$$(3.6) \qquad \dot{\varphi} = -(\sin\varphi + k) \ .$$

No stationary point exists if $k > 1$, i.e., $l > R$. For $k < 1$ the system has two stationary points φ_1 and φ_2 where $-\pi/2 \leq \varphi_1 \leq 0$ and $\varphi_2 = 2\pi - \varphi_1$. If $|\varphi_1| < \varphi_m < |\varphi_2|$, then φ_1 is a global attractor. If $\varphi_m < |\varphi_1|$ the vehicle will get stuck at a certain moment. The cases $k = 1$ and $\varphi_1 = \varphi_2 = -\pi/2$ deserve special attention. Clearly, in this case we assume that $\varphi_m \geq \frac{\pi}{2}$. In the stationary situation the rear wheel remains in the centre of the circle while the front wheel turns around. The corresponding picture of the phase space is given in Fig. 2. Note, that $\varphi = -\pi/2$ is a global attractor only if $\varphi_m = \infty$. It is not a stable point in the sense of Lyapunov (see, e.g., reference [5]), because an orbit starting at $\varphi_0 = -\pi/2-\varepsilon$ ($\varepsilon > 0$) will always reach $\varphi = +\pi/2$ regardless of the value of ε. The vehicle completely turns around its own front wheel and then approaches the stationary situation $\varphi = -\pi/2$.

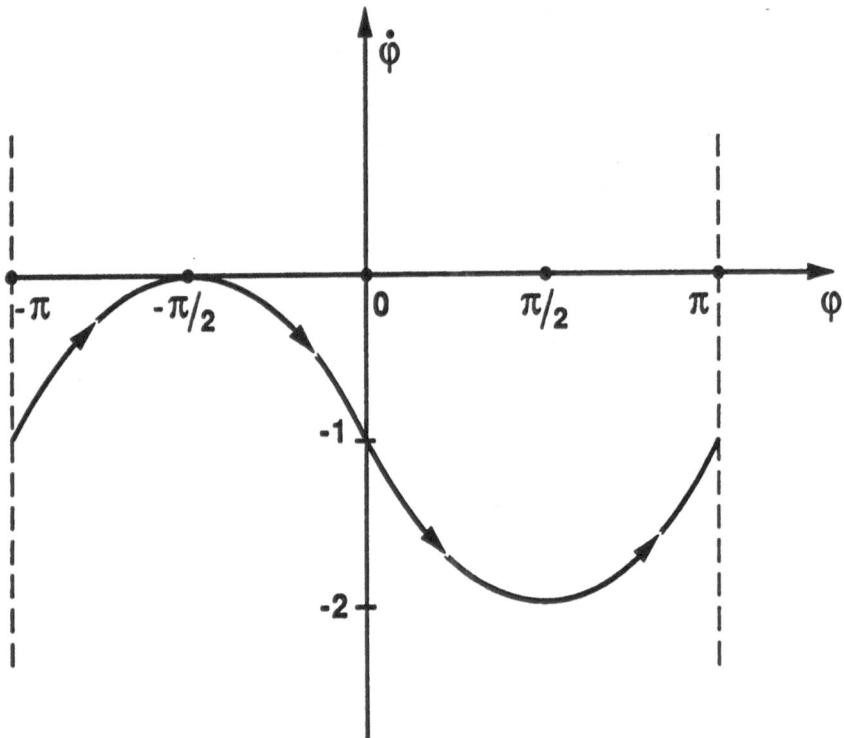

Fig. 2. Faseplane for $\alpha = 0$, $k = 1$. The vehicle follows the curve drawn.

Turning

We now consider the effect of rear-wheel steering while the vehicle is turning. For the path of the front wheel we take parts of a straight line, a circle and again a straight line respectively. This path is drawn in Fig. 3. At $t = 0$ we take $\varphi_0 = 0$ and the vehicle just enters the turn. The radius of the (circular) turn is R, measured in units of l. As mentioned earlier, the dimensionless parameter t is equal to the arc length along the front wheel path if the velocity is constant. This means that the front wheel has finished turning at $t = \pi R/2$. During turning the angle $\varphi(t)$ is given by

$$(3.7a) \quad \varphi(t) = \int_0^t f(\varphi(\tau)) \, d\tau$$

where

$$(3.7b) \quad f(\varphi) = \begin{cases} -(\sin \varphi - \tan \alpha\varphi \, \cos \varphi + 1/R) & ; \ 0 \le t < \pi R/2 \\ -(\sin \varphi - \tan \alpha\varphi \, \cos \varphi) & ; \ t \ge \pi R/2 \, . \end{cases}$$

The position $\mathbf{x}_A(t)$ of the front wheel is given explicitly by

$$\mathbf{x}_A(t) = \begin{cases} R \begin{pmatrix} \sin(t/R) \\ -\cos(t/R) \end{pmatrix} & ; \ 0 \le t < \pi R/2 \\ \begin{pmatrix} R \\ t - (\pi R/2) \end{pmatrix} & ; \ t \ge \pi R/2 \, . \end{cases}$$

For $0 \le t < \pi R/2$ the basis vectors $(\mathbf{t}, \mathbf{n})$ are represented by

$$\mathbf{t}(t) = \begin{pmatrix} \cos(t/R) \\ \sin(t/R) \end{pmatrix}$$

$$\mathbf{n}(t) = \begin{pmatrix} -\sin(t/R) \\ \cos(t/R) \end{pmatrix} ,$$

and for $t \ge \pi R/2$ by

$$\mathbf{t}(t) = \begin{pmatrix} 0 \\ 1 \end{pmatrix}$$

$$\mathbf{n}(t) = \begin{pmatrix} -1 \\ 0 \end{pmatrix} .$$

The position $\mathbf{x}_B(t)$ of the rear wheel follows from

$$\mathbf{x}_B(t) = \mathbf{x}_A(t) - (\cos \varphi(t) \, \mathbf{t}(t) + \sin \varphi(t) \, \mathbf{n}(t)) \, .$$

Fig. 3 illustrates the trajectory $x_B(t)$, $-1 \leq t \leq 2.3\pi R$ for the cases $\alpha = -0.1$, 0.0, and 0.3. For convenience we set $R = 1.0$. The numerical integration in (3.7a) is performed with a Runge-Kutta procedure. From Fig. 3 it is clear that the trajectory of the rear wheel is quite sensitive to the value of α. A small, negative value of $\alpha(\alpha \approx -0.1)$ has already the effect that the trajectories of front and rear wheels almost coincide.

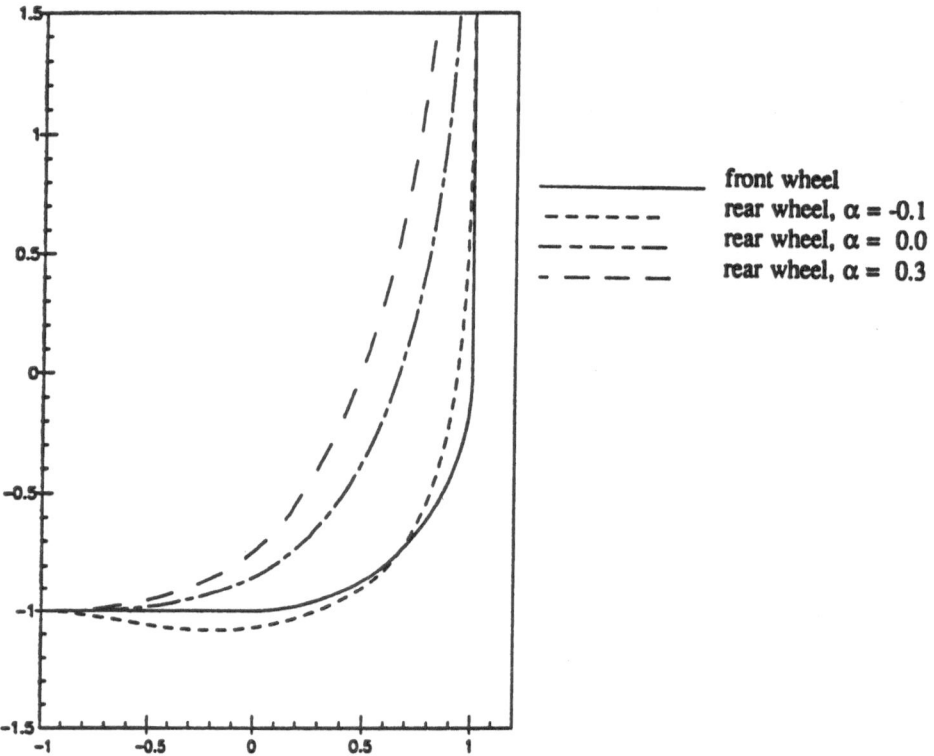

Fig. 3. The paths of the front wheel (solid line) and the rear wheel (dotted lines) while taking a turn of 90 degrees.

We shall not work out here modeling steps 6: *Verification*, 7: *Reiteration*, and 8: *Implementation*. The verification would imply that we compare measured and calculated trajectories. It might appear that the model describes the experiments very well, but only if the width of the vehicle is small compared to the radius of curvature of the path followed. One might come to the conclusion that it is absolutely necessary to take into account that real trucks have not two wheels but four or more. These details would require an extension of model (3.4). The implementation might vary from customer to customer. It will always comprise a computer program, in which the differential equations are numerically solved. The set-up of the input and output of this program can only be organized effectively in close interaction with the customer.

References

Here, we list some introductory texts on the topics of this chapter. If a book is of an advanced level this is denoted by an (∗).

On mathematical modeling:

1. F.A. BENDER, *An Introduction to Mathematical Modeling*, John Wiley & Sons, New York, 1978, ISBN 0–471–02951–3.

2. J.S. BERRY, D.N. BURGHES, I.D. HUNTLEY (eds.), *Teaching and Applying Mathematical Models*, Ellis Horwood, Chichester, 1985, ISBN 0–85312–728–X.

3. D.N. BURGHES, M.S. BORRIE, Chichester, Ellis Horwood, 1981.

4. R.R. CLEMENTS, *Mathematical Modeling, a Case Study Approach*, Cambridge University Press, 1989, ISBN 0–521–34340–2.

5. D.W. JORDAN, P. SMITH, *Non-linear Ordinary Differential Equations*, Oxford University Press, 1977, (∗).

6. A.B. TAYLER, *Mathematical Models in Applied Mechanics*, Clarendon Press, Oxford, 1986, ISBN 0–19–853533–3.

On dimensional analysis:

7. G.I. BARENBLATT, *Similarity, Self-Similarity, and Intermediate Asymptotics*, Consultants Bureau, New York, 1979, ISBN 0–306–10956–5.

8. E.A. BENDER, see [1].

On similarity solutions:

9. G.I. BARENBLATT, see [7].

10. A.G. HANSEN, *Similarity Analyses of Boundary Value Problems in Engineering*, Prentice-Hall, London, 1964.

On differential geometry:

11. A.R. FORSYTH, *Lectures on the Differential Geometry of Curves and Surfaces*, Cambridge University Press, 1912.

12. C.E. WEATHERBURN, *Differential Geometry of Three Dimensions*, Cambridge University Press, 1927, (*).

13. D.J. STRUIK, *Lectures on Classical Differential Geometry*, Addison-Wesley, London, 1957, (*).

General Handbooks:

14. A.M. ABRAMOWITZ, I.A. STEGUN, *Handbook of Mathematical Functions*, Dover, 1965.

15. I.S. GRADSHTEYN, I.M. RYZHIK, *Tables of Integrals, Series and Products*, Academic Press, 1965.

About difference equations, algebras and discrete events

G.J. Olsder

Faculty of Technical Mathematics and Informatics

Delft University of Technology

P.O. Box 5031, 2600 GA Delft, The Netherlands

Abstract

An introduction to the theory of discrete event dynamic systems is given. Discrete event dynamic systems (DEDS) are nonlinear in the conventional algebra, but are linear in the max-plus algebra. Of many concepts and results within the conventional linear algebra and linear systems theory duplicates exist in the max-plus algebra and the theory of DEDS. The motivation to study DEDS comes from the description of flows in networks. Such networks are for instance related to computer systems, traffic systems and flexible manufacturing in production planning.

1 Difference Equations

1.1 Introduction

A well known equation in the theory of difference equations is the linear equation

$$x(t+1) = Ax(t), \quad t = 0, 1, 2, \ldots \tag{1}$$

The vector $x \in R^n$ represents the 'state' of an underlying model and this state evolves in time according to this equation; $x(t)$ denotes the state at time instant t. The symbol A represents a given $n \times n$ matrix. If an initial condition

$$x(0) = x_0 \tag{2}$$

is given, then the whole future evolution of (1) is determined.

A. van der Burgh and J. Simonis (eds.), Topics in Engineering Mathematics, 121–150.

Implicit in the text above is that (1) is a vector equation. Written out in scalar equations it becomes

$$x_i(t+1) = \sum_{j=1}^{n} a_{ij} x_j(t), \quad i = 1, \ldots, n; \quad t = 0, 1, \ldots \qquad (3)$$

The symbol x_i denotes the i-th component of the vector x; the elements a_{ij} are the entries of the square matrix A. If $a_{ij}, i, j = 1, \ldots, n$ and $x_j(t), j = 1, \ldots, n$ are given, then $x_j(t+1), j = 1, \ldots, n$, can be calculated according to (1) or (3).

As an example take $n = 2$, such that A is a 2×2 matrix. Take

$$A = \begin{pmatrix} 3 & 7 \\ 2 & 4 \end{pmatrix} \qquad (4)$$

and as initial condition

$$x_0 = \begin{pmatrix} 1 \\ 0 \end{pmatrix} . \qquad (5)$$

The time evolution of (1) becomes for this example

$$x(0) = \begin{pmatrix} 1 \\ 0 \end{pmatrix}, \quad x(1) = \begin{pmatrix} 3 \\ 2 \end{pmatrix}, \quad x(2) = \begin{pmatrix} 23 \\ 14 \end{pmatrix}, \quad x(3) = \begin{pmatrix} 167 \\ 102 \end{pmatrix}, \ldots \qquad (6)$$

1.2   Solution by means of Eigenvectors

Assume that the initial vector (2) equals an eigenvector of A; the corresponding eigenvalue is denoted by λ. The solution of (1) can be written as

$$x(t) = \lambda^t x_0, \quad t = 0, 1, \ldots \qquad (7)$$

More generally, if the initial vector can be written as a linear combination of the set of linearly independent eigenvectors;

$$x_0 = \sum_j c_j v_j , \qquad (8)$$

where v_j is the j-th eigenvector with corresponding eigenvalue λ_j, the c_j are coefficients, then

$$x(t) = \sum_j c_j \lambda_j^t v_j .$$

If the matrix A is diagonizable, then the set of linearly independent eigenvectors spans R^n, and any initial condition x_0 can be expressed as in (8). If A is not diagonizable, then one must work with generalized eigenvectors and the formula which expresses $x(t)$ in terms of eigenvalues and x_0 is slightly more complicated. This complication will not occur in the current context and therefore will not be dealt with explicitly.

2 Changing the Algebra

2.1 The Max-Plus Algebra

The only operations used in (1) or (3) are multiplication $(a_{ij} \times x_j(t))$ and addition (the $\sum$ symbol). Most of this paper can be considered as a study of formulas of the form (1), in which the operations are changed. Suppose that the two operations in (3) are changed in the following way; addition becomes maximization and multiplication becomes addition. Then (3) becomes

$$
\begin{aligned}
x_i(k+1) &= \max(a_{i1} + x_1(k), a_{i2} + x_2(k), \ldots, a_{in} + x_n(k)) \\
&= \max_j(a_{ij} + x_j(k)), \quad i = 1, \ldots, n \ .
\end{aligned}
\tag{9}
$$

If the initial condition (2) also holds for (9), then the time evolution of (9) is completely determined again. Of course the time evolutions of (3) and (9) will be different in general. Equation (9), as it stands, is a nonlinear difference equation. As an example take A from (4) and $x(0)$ from (5). Then the time evolution of (9) becomes

$$
x(0) = \begin{pmatrix} 1 \\ 0 \end{pmatrix}, \ x(1) = \begin{pmatrix} 7 \\ 4 \end{pmatrix}, \ x(2) = \begin{pmatrix} 11 \\ 9 \end{pmatrix}, \ x(3) = \begin{pmatrix} 16 \\ 13 \end{pmatrix}, \ldots \tag{10}
$$

2.2 Motivation

We are used to thinking of the argument t in $x(t)$ as a time instant; at time instant t the state is $x(t)$. With respect to (9) we will introduce a different meaning for this argument. In order to emphasize this different meaning, the argument t has already been replaced by k. For a practical motivation we need to think of a network, which consists of a number of nodes and some arcs connecting these nodes. The network corresponding to (9) has n nodes; one for each component x_i. Entry a_{ij} corresponds to the arc from node j to node i. In terms of graph theory such a network is called a directed graph ('directed' because the individual arcs between the nodes are one way

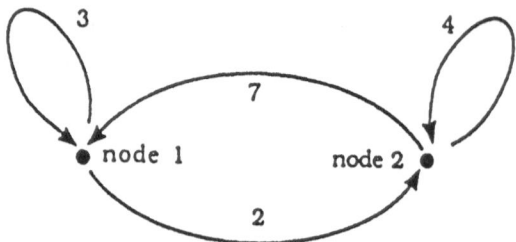

Figure 1: Network corresponding to Equation (4)

arrows). Therefore the arcs corresponding to a_{ij} and a_{ji}, if both exist, are considered to be different.

The nodes in the network can perform certain activities; each node has its own kind of activity. Such activities take a finite time, called holding time, to be performed. These holding times may be different for different nodes. It is assumed that an activity at a certain node can only start when all preceding ('directly upstream') nodes have finished their activities and sent the results of these activities along the arcs to the current node. Thus the arc corresponding to a_{ij} can be interpreted as an output channel for node j and simultaneously as an input channel for node i. Suppose that this node i starts its activity as soon as all preceding nodes have sent their results (the rather neutral word 'results' is used, it could equally have been messages, ingredients or products,...) to node i, then (9) describes when the activities take place. The interpretation of the quantities used is:

- $x_i(k)$: is the earliest time instant at which node i becomes active for the k-th time;

- a_{ij} : is the sum of the holding time (i.e. time duration of the activity) at node j and the travelling time (the rather neutral 'travelling time' is used rather than for instance 'transportation time' or 'communication time') from node j to node i.

The fact that we write a_{ij} rather than a_{ji} for a quantity connected to the arc from node j to node i has to do with matrix equations which will be written in the classical way with column vectors, as will be seen later on. (This is in contrast with queuing theory, where it is customary to work with row vectors.) For the example given above, the network has two nodes and four arcs, as given in Figure 1. The interpretation of the number 3 in this figure is that if node 1 has started an activity, the next activity cannot start within the next 3 time units. Similarly, the time between two subsequent activities of node 2 is at least 4 time units. Node 1 sends its results to node 2 and once an activity starts in node 1, it takes 2 time units before the result of this activity reaches node 2. Similarly it takes 7 time units after the initiation of an activity of node 2 for the result of that activity to reach node 1. Suppose that an activity refers to some production. The production time of node 1 could for instance be 1 unit of time; after that, node 1 needs 2 time units for recovery (lubrication say) and the travelling time of the result (the final product) from node 1 to node 2 is 1 unit of time. Thus the number $a_{11} = 3$ is made up of a production time 1 and a recovery time 2 and the number $a_{21} = 2$ is made up of the same production time 1 and a travelling time 1. Similarly, if the production time at node 2 is 4, the this node does not need any time for recovery (because $a_{22} = 4$), and the travelling time from node 2 to node 1 is 3 (because $a_{12} = 7 = 4 + 3$).

If we now look at the sequence (10) again, the interpretation of the vectors $x(k)$ is different from the initial one. The argument k is not a time instant anymore, but a counter which states how many times the various nodes have been active. At time 14 node 1 has been active twice (more precisely, node 1 has started two activities, respectively at times 7 and 11). At the same time 14, node 2 has been active three times (it started activities at times 4, 9 and 13). The counting of the activities is such that it coincides with the argument of the x vector. The initial condition is henceforth considered to be the 0-th activity.

In Figure 1 there was an arc from any node to any other node. In many networks referring to more practical situations, this will not be the case. If there is no arc from node j to node i then node i does not need any result from node j. Therefore node j does not have a direct influence on the behavior of node i. In such a situation it is useful to consider the element a_{ij} to be equal to $-\infty$. In (9) a term $-\infty + x_j(k)$ does not influence $x_i(k+1)$ as long as $x_j(k)$ is finite. The number $-\infty$ will occur frequently in the sequel and it will be indicated by ε.

2.3   Some Notation and Some Calculus

For reasons which will become clear later on, (9) will be written as

$$x_i(k + 1) = \bigoplus_j a_{ij} \otimes x_j(k), \quad i = 1, \ldots, n ,$$

or in vector notation,

$$x(k+1) = A \otimes x(k) . \qquad (11)$$

The symbol $\bigoplus_j c(j)$ refers to the maximum of the elements $c(j)$ with respect to all appropriate j, and $\otimes$ refers to addition. Later on the symbol $\oplus$ will also be used; $a \oplus b$ refers to the maximum of the scalars a and b. If the initial condition for (11) is $x(0) = x_0$, then

$$x(1) = A \otimes x_0 ,$$

$$x(2) = A \otimes x(1) = A \otimes (A \otimes x_0) = (A \otimes A) \otimes x_0 = A^2 \otimes x_0 .$$

It can be shown that indeed $A \otimes (A \otimes x_0) = (A \otimes A) \otimes x_0$. For the example given above it is easy to check this by hand. Instead of $A \otimes A$ we simply write A^2. We get

$$x(3) = A \otimes x(2) = A \otimes (A^2 \otimes x_0) = (A \otimes A^2) \otimes x_0 = A^3 \otimes x_0 ,$$

and in general

$$x(k) = (\underbrace{A \otimes A \otimes \cdots \otimes A}_{k \text{ times}}) \otimes x_0 = A^k \otimes x_0 .$$

The matrices A^2, A^3,..., can be calculated directly. Let us consider the A-matrix of (4) again, then

$$A^2 = \begin{pmatrix} \max(3+3, 7+2) & \max(3+7, 7+4) \\ \max(2+3, 4+2) & \max(2+7, 4+4) \end{pmatrix} = \begin{pmatrix} 9 & 11 \\ 6 & 9 \end{pmatrix} .$$

In general

$$(A^2)_{ij} = \bigoplus_l a_{il} \otimes a_{lj} = \max_l(a_{il} + a_{lj}) . \qquad (12)$$

The quantity $(A^2)_{ij}$ can be interpreted as the maximum (with respect to l) of all connections from node j via node l to node i. One speaks of paths of length two between the nodes j and i. More generally, $(A^k)_{ij}$ denotes the maximum of all paths of length k, starting at node j and ending at node i.

The multiplication of two matrices in the max-plus algebra follows the standard pattern as shown by the example

$$\begin{pmatrix} 1 & 2 \\ \varepsilon & 0 \\ -2 & 1 \end{pmatrix} \begin{pmatrix} 5 & 2 \\ 1 & 3 \end{pmatrix} = \begin{pmatrix} 6 & 5 \\ 1 & 3 \\ 3 & 4 \end{pmatrix} .$$

2.4 Axiomatics

The operations $\oplus$ and $\otimes$ defined on the set R can also be defined with respect to a more general set of elements $\mathcal{D}$. One then speaks of a *dioid*.

Definition .1 (Dioid) *A dioid is a set $\mathcal{D}$ endowed with two operations denoted $\oplus$ and $\otimes$ (called 'sum' or 'addition', and 'product' or 'multiplication') obeying the following axioms:*

Axiom .2 (Associativity of addition)

$$\forall a, b, c \in \mathcal{D}, (a \oplus b) \oplus c = a \oplus (b \oplus c) \ .$$

Axiom .3 (Commutativity of addition)

$$\forall a, b \in \mathcal{D}, a \oplus b = b \oplus a \ .$$

Axiom .4 (Associativity of multiplication)

$$\forall a, b, c \in \mathcal{D}, (a \otimes b) \otimes c = a \otimes (b \otimes c) \ .$$

Axiom .5 (Distributivity)

$$\forall a, b, c \in \mathcal{D}, \quad (a \oplus b) \otimes c = (a \otimes c) \oplus (b \otimes c) \ ,$$
$$c \otimes (a \oplus b) = c \otimes a \oplus c \otimes b \ .$$

This is right, respectively left, distributivity of multiplication with respect to addition. One statement does not follow from the other since multiplication is not assumed to be commutative.

Axiom .6 (Existence of a zero element)

$$\exists \varepsilon \in \mathcal{D} : \forall a \in \mathcal{D}, a \oplus \varepsilon = a \ .$$

Axiom .7 (Absorbing zero element)

$$\forall a \in \mathcal{D}, a \otimes \varepsilon = \varepsilon \otimes a = \varepsilon \ .$$

Axiom .8 (Existence of an identity element)

$$\exists e \in \mathcal{D} : \forall a \in \mathcal{D}, a \otimes e = e \otimes a = a \ .$$

Axiom .9 (Idempotency of addition)

$$\forall a \in \mathcal{D}, a \oplus a = a \ .$$

Definition .10 (Commutative dioid) *A dioid is* commutative *if multiplication is commutative.*

With the noticeable exception of Axiom .9, most the axioms of dioids are required for rings too. Indeed, Axiom .9 is the most distinguishing feature of dioids. Because of this axiom, addition cannot be cancellative, that is, $a \oplus b = a \oplus c$ does not imply $b = c$ in general. Multiplication is not necessarily cancellative either (of course, because of Axiom .7, cancellation would anyway only apply to elements different from ε). For an example in which multiplication is not cancellative take $\mathcal{D} = R \cup \{-\infty\} \cup \{+\infty\}$ and define $\oplus$ as max and $\otimes$ as min.

It is easily shown that in dioids the distributivity with respect to matrices also holds, i.e. $A \otimes (B \otimes C) = (A \otimes B) \otimes C$, where these multiplications only make sense if the matrices have appropriate dimensions.

2.5 Systems with Inputs and Outputs

An extension of (11) is

$$\left. \begin{array}{rcl} x(k+1) & = & (A \otimes x(k)) \oplus (B \otimes u(k)) \ , \\ y(k) & = & C \otimes x(k) \ . \end{array} \right\} \tag{13}$$

The symbol $\oplus$ in this formula refers to componentwise maximization. The m-vector u is called the input to the system; the p-vector y is the output of the system. The components of u refer to nodes which have no predecessors. Similarly, the components of y refer to nodes with no successors. The components of x now refer to internal nodes, i.e. to nodes with have both successors and predecessors. The matrices $B = \{b_{ij}\}$ and $C = \{c_{ij}\}$ have sizes $n \times m$ and $p \times n$ respectively. The traditional way of writing (13) would be

$$\begin{array}{rcl} x_i(k+1) & = & \max(a_{i1} + x_1(k), \ldots, a_{in} + x_n(k), \\ & & \quad b_{i1} + u_1(k), \ldots, b_{im} + u_m(k)), \quad i = 1, \ldots, n \ ; \\ y_i(k) & = & \max(c_{i1} + x_1(k), \ldots, c_{in} + x_n(k)), \quad i = 1, \ldots, p \ . \end{array}$$

Usually (13) is written as

$$\left. \begin{array}{rcl} x(k+1) & = & Ax(k) \oplus Bu(k) \ , \\ y(k) & = & Cx(k) \ . \end{array} \right\}, \tag{14}$$

where it is understood that multiplication has priority over addition. If it is clear where the '$\otimes$'-symbols are used, they are sometimes omitted, as shown

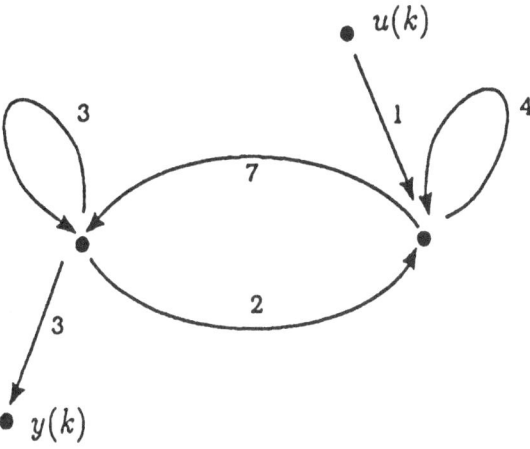

Figure 2: Network with input and output

in (14). This practice is exactly the same one as with respect to the more common multiplication ' × ' or ' . ' symbol in conventional algebra. In the same vein, in conventional algebra $1 \times x$ is the same as $1x$, which is usually written as x. Within the context of the $\otimes$ and $\oplus$ symbols, $0 \otimes x$ is exactly the same as x. The symbol ε is the neutral element with respect to maximization; its numerical value equals $-\infty$. Similarly, the symbol e will denote the neutral element with respect to addition; it assumes the numerical value 0. Note that $1 \otimes x$ is different from x.

If one wants to think in terms of a network again, then $u(k)$ is a vector indicating when certain resources become available for the k-th time. Subsequently it takes b_{ij} time units before the j-th resource reaches node i of the network. The vector $y(k)$ refers to the time instant at which the final products of the network are delivered to the outside world.

Take for example

$$x(k+1) = \begin{pmatrix} 3 & 7 \\ 2 & 4 \end{pmatrix} x(k) \oplus \begin{pmatrix} \varepsilon \\ 1 \end{pmatrix} u(k) ,$$

$$y(k) = (3 \quad \varepsilon)x(k) .$$

(15)

The corresponding network is shown in Figure 2. Because $b_{11} = \varepsilon(= -\infty)$, the input $u(k)$ only goes to node 2. If one would replace B by $(2, \quad 1)'$ for instance, then each input would 'spread' itself over the two nodes. In this example with $B = (2, \quad 1)'$, from time instant $u(k)$ on, it takes 2 time units for the input to reach node 1 and 1 time unit to reach node 2. In many practical situations an input will enter the network through one node. That is why in (15) only one b_i-component is different from ε. Similar remarks can be made with respect to the output. Suppose that we have (5) as an initial condition and that

$$u(0) = 1, u(1) = 7, u(2) = 13, u(3) = 19, \ldots,$$

then it easily follows that

$$x(0) = \begin{pmatrix} 1 \\ 0 \end{pmatrix}, x(1) = \begin{pmatrix} 7 \\ 4 \end{pmatrix}, x(2) = \begin{pmatrix} 11 \\ 9 \end{pmatrix}, x(3) = \begin{pmatrix} 16 \\ 14 \end{pmatrix}, \ldots$$

$$y(0) = 4, y(1) = 10, y(2) = 14, y(3) = 19, \ldots$$

2.6   Higher Order Difference Equations

We started this section with the difference equation (1), which is a first order linear vector difference equation. It is well known that a higher order linear scalar difference equation

$$z(k+1) = a_1 z(k) + a_2 z(k-1) + \cdots + a_n z(k-n+1) \qquad (16)$$

can be written in the form of equation (1). If we introduce the vector $(z(k), z(k-1), \ldots, z(k-n+1))'$, then (16) can be written as

$$\begin{pmatrix} z(k+1) \\ z(k) \\ \vdots \\ \vdots \\ z(k-n+2) \end{pmatrix} = \begin{pmatrix} a_1 & a_2 & \cdots & \cdots & a_n \\ 1 & 0 & \cdots & \cdots & 0 \\ 0 & & & & \\ \vdots & & & & \\ 0 & \cdots & 0 & 1 & 0 \end{pmatrix} \begin{pmatrix} z(k) \\ z(k-1) \\ \vdots \\ \vdots \\ z(k-n+1) \end{pmatrix}. \qquad (17)$$

This equation has exactly the form of (1). If we change the operations in (16) in the standard way; addition becomes maximization and multiplication becomes addition, then the numerical evaluation of (16) becomes

$$z(k+1) = \max(a_1 + z(k), a_2 + z(k-1), \ldots, a_n + z(k-n+1)) . \qquad (18)$$

This equation can also be written as a first order linear vector difference equation. In fact this equation is almost Equation (17), which must now be evaluated with the operations maximization and addition. The only difference is that the 1's and 0's in the matrix in (17) must be replaced by e's and ε's respectively.

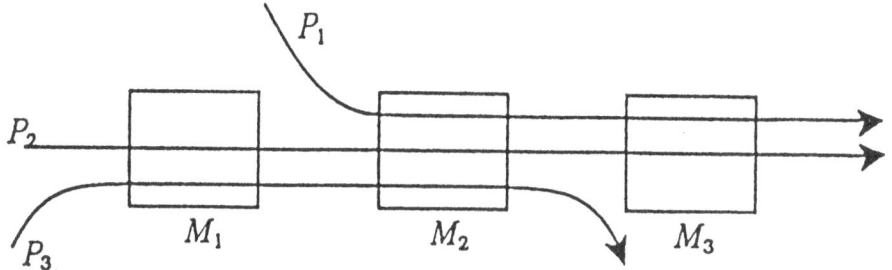

Figure 3: Routing of parts along machines

Table 1: Processing times

	P_1	P_2	P_3
M_1		1	5
M_2	3	2	3
M_3	4	3	

3   Example on Production

Consider a manufacturing system consisting of three machines. It is supposed to produce three kinds of parts according to a certain product mix.
The routes to be followed by each part and each machine are depicted in
Figure 3; in which $M_i, i = 1, 2, 3$, are the machines and $P_i, i = 1, 2, 3$, are the
parts. Processing times are given in Table 1. Note that this manufacturing
system has a flow-shop structure, i.e. all parts follow the same sequence on
the machines (although they may skip some) and every machine is visited
at most once by each part. This manufacturing system is automated and
there are no set-up times on machines when they switch from one part type
to another. Parts are carried on a limited number of pallets (or, equivalently, product carriers) by means of fixtures. For reasons of simplicity it is
assumed that

1. only one pallet is available for each part type;

2. there are no set-up times or travelling times;

3. the sequencing of part types on the machines is known and it is
   (P_2, P_3) on M_1, (P_1, P_2, P_3) on M_2 and (P_1, P_2) on M_3.

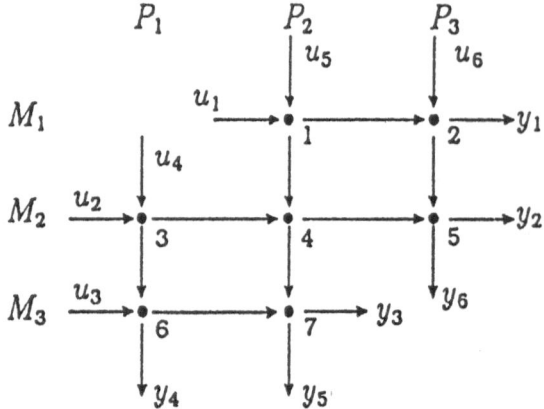

Figure 4: The ordering of activities in the flexible manufacturing system

The last point mentioned is not for reasons of simplicity. If any machine would start working on the part which would arrive first instead of waiting for the appropriate part, the modelling cannot be done within the context of the max-plus algebra. In order to accomodate the rule 'first in, first served', one would need a third operation, viz. min and that would make the general modeling more difficult and less transparant.

We can draw a graph in which each node corresponds to a combination of a machine and a part. Since M_1 works on 2 parts, M_2 on 3 and M_3 on 2, this graph has seven nodes. The arcs between the nodes express the precedence constraints between operations due to the sequencing of operations on the machines. To each node i in Figure 4 corresponds a number x_i; it denotes the earliest time instant at which the node can start its activity. In order to be able to calculate these quantities, the time instants at which the machines and parts (together called the resources) are available must be given. This is done by means of a six-dimensional input vector u (six since there are six resources: three machines and three parts). There is an output vector also; the elements of the six-dimensional vector y denote the time instants at which the parts are ready and the machines have finished their jobs (for one cycle). The model becomes

$$x = Ax \oplus Bu \; ; \tag{19}$$

$$y = Cx \; , \tag{20}$$

in which the matrices are

$$
A = \begin{pmatrix}
\varepsilon & \varepsilon & \varepsilon & \varepsilon & \varepsilon & \varepsilon & \varepsilon \\
1 & \varepsilon & \varepsilon & \varepsilon & \varepsilon & \varepsilon & \varepsilon \\
\varepsilon & \varepsilon & \varepsilon & \varepsilon & \varepsilon & \varepsilon & \varepsilon \\
1 & \varepsilon & 3 & \varepsilon & \varepsilon & \varepsilon & \varepsilon \\
\varepsilon & 5 & \varepsilon & 2 & \varepsilon & \varepsilon & \varepsilon \\
\varepsilon & \varepsilon & 3 & \varepsilon & \varepsilon & \varepsilon & \varepsilon \\
\varepsilon & \varepsilon & \varepsilon & 2 & \varepsilon & 4 & \varepsilon
\end{pmatrix} ; \quad
B = \begin{pmatrix}
e & \varepsilon & \varepsilon & \varepsilon & e & \varepsilon \\
\varepsilon & \varepsilon & \varepsilon & \varepsilon & \varepsilon & e \\
\varepsilon & e & \varepsilon & e & \varepsilon & \varepsilon \\
\varepsilon & \varepsilon & \varepsilon & \varepsilon & \varepsilon & \varepsilon \\
\varepsilon & \varepsilon & \varepsilon & \varepsilon & \varepsilon & \varepsilon \\
\varepsilon & \varepsilon & e & \varepsilon & \varepsilon & \varepsilon \\
\varepsilon & \varepsilon & \varepsilon & \varepsilon & \varepsilon & \varepsilon
\end{pmatrix} ;
$$

$$
C = \begin{pmatrix}
\varepsilon & 5 & \varepsilon & \varepsilon & \varepsilon & \varepsilon & \varepsilon \\
\varepsilon & \varepsilon & \varepsilon & \varepsilon & 3 & \varepsilon & \varepsilon \\
\varepsilon & \varepsilon & \varepsilon & \varepsilon & \varepsilon & \varepsilon & 3 \\
\varepsilon & \varepsilon & \varepsilon & \varepsilon & \varepsilon & 4 & \varepsilon \\
\varepsilon & \varepsilon & \varepsilon & \varepsilon & \varepsilon & \varepsilon & 3 \\
\varepsilon & \varepsilon & \varepsilon & \varepsilon & 3 & \varepsilon & \varepsilon
\end{pmatrix} .
$$

$$(21)$$

Equation (19) is an implicit equation in x. Let us see what we get by repeated substitution of the complete right-hand side of (19) for x of this same right-hand side. After one substitution:

$$
\begin{aligned}
x &= A^2 x \oplus ABu \oplus Bu \\
&= A^2 x \oplus (A \oplus e)Bu \ ,
\end{aligned}
$$

and after k substitutions:

$$
x = A^k x \oplus (A^{k-1} \oplus A^{k-2} \oplus \cdots \oplus A \oplus e)Bu \ .
$$

In the formulas above e refers to the identity matrix; zeros on the diagonal and ε's elsewhere. The symbol e will be used as the identity element for all spaces that will be encountered in this paper. Similarly, ε will be used to denote the zero element of any space to be encountered.

For this example it is easily shown that $A^n = \varepsilon = -\infty$ for $n \geq 3$. In the next session a graph theoretic explanation for these equalities will be given (the precedence graph of A is acyclic). Therefore the solution x in the current example becomes

$$
x = (A^2 \oplus A \oplus e)Bu \ ,
$$

for which we can write

$$
x = A^* Bu \ ,
$$

where A^* is defined as

$$A^* \overset{\text{def}}{=} e \oplus A \oplus \cdots \oplus A^n \oplus A^{n+1} \oplus \cdots. \tag{22}$$

In our example on production, $A^k, k > 2$, does not contribute to the sum in (22). For later reference, we also introduce the notation

$$A^+ \overset{\text{def}}{=} A \oplus \cdots \oplus A^n \oplus A^{n+1} \oplus \cdots. \tag{23}$$

Remark .11 With the conventional matrix calculus in mind one might be tempted to write for (22):

$$(e \oplus A \oplus A^2 \oplus \cdots) = (e \ominus A)^{-1} . \tag{24}$$

Of course, we have not defined the inverse of a matrix within the current setting and (24) is an empty statement. It is also strange to have a 'minus' sign $\ominus$ in (24) and it is not known how to interpret this sign in the context of the max-operation at the left-hand side of the equation. It should be the reverse operation of $\oplus$. If we dare to continue along these shaky lines, one could calculate the solution of (19) as

$$(e \ominus A)x = Bu \Rightarrow x = (e \ominus A)^{-1}Bu ,$$

which equals $x = A^*Bu$ if we believe (24) to make sense. Quite often one can guide one's intuition by considering formal expressions of the kind (24). One tries to find formal analogies in the notation with the conventional analysis. It can be shown [1] that an inverse as in (24) does not exist in general and therefore we get 'stuck' with the series expansion. ∎

Now we add feedback arcs to Figure 4 as illustrated in Figure 5. In this graph the feedback arcs are indicated by dotted lines. The meaning of these feedback arcs is the following. After a machine has finished a sequence of products, it starts with the next sequence. If the pallet on which product P_i was mounted is at the end, the finished product is removed and the empty pallet immediately goes back to the starting point to pick up a new part P_i. If it is assumed that the feedback arcs have zero time duration, then $u(k) = y(k-1)$, where $u(k)$ is the k-th input cycle and $y(k)$ the k-th output. Thus we can write

$$\begin{aligned} y(k) &= Cx(k) &= CA^*Bu(k) \\ &&= CA^*By(k-1) . \end{aligned} \tag{25}$$

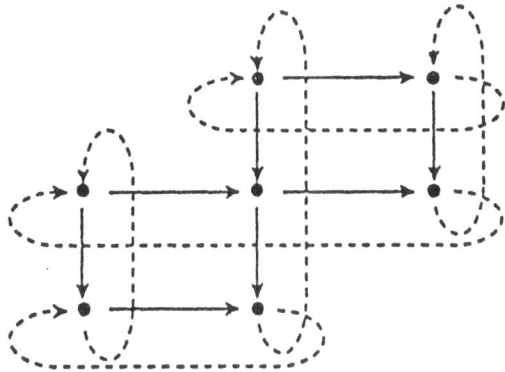

Figure 5: Production system with feedback arcs

The transition matrix from $y(k-1)$ to $y(k)$ can be calculated (a simple computer program does the job, but it can be done by hand);

$$M \stackrel{\text{def}}{=} CA^*B = \begin{pmatrix} 6 & \varepsilon & \varepsilon & \varepsilon & 6 & 5 \\ 9 & 8 & \varepsilon & 8 & 9 & 8 \\ 6 & 10 & 7 & 10 & 6 & \varepsilon \\ \varepsilon & 7 & 4 & 7 & \varepsilon & \varepsilon \\ 6 & 10 & 7 & 10 & 6 & \varepsilon \\ 9 & 8 & \varepsilon & 8 & 9 & 8 \end{pmatrix}. \tag{26}$$

This matrix M determines the speed with which the manufacturing system can work. We will come back to this issue in the next section.

4 Some Graph Theory and the Spectral Theory of Matrices

4.1 Graph Theory

Informally, we already encountered several items directly related to graphs. We will formalize some of these concepts here. A *directed graph* $\mathcal{G}$ is defined as a pair $(\mathcal{V}, \mathcal{E})$, where $\mathcal{V}$ is a set of elements called *nodes* and where $\mathcal{E}$ is a set of which the elements are ordered (not necessarily different) pairs of nodes, called *arcs*. The possibility of several arcs between two nodes exists (one then speaks about a multigraph); in this paper, however, we exclusively deal with directed graphs in which there is at most one (i.e. zero or one) arc between any two nodes. Instead of directed graph one often uses the shorter word 'digraph', or even 'graph' if it is clear from the context that digraph is meant.

Denote the number of nodes by n, and number the individual nodes $1, 2, \ldots, n$. If $(i,j) \in \mathcal{E}$, then i is called the initial node or the origin of the arc (i,j), and j the final node or the destination of the arc (i,j). Graphically, the nodes are represented by points, and the arc (i,j) is represented by an 'arrow' from i to j.

We now give a list of some concepts of graph theory which will be used later on.

Predecessor, successor. If in a graph $(i,j) \in \mathcal{E}$ then i is called a predecessor of j and j is called a successor of i. The set of all predecessors of j is indicated by $\pi(j)$ and the set of all successors of i is indicated by $\sigma(i)$. A predecessor is also called an *upstream node* and a successor is also called a *downstream node*.

Source, sink. If $\pi(i) = \emptyset$ then node i is called a source; if $\sigma(i) = \emptyset$ then i is called a sink. Depending on the application, a source, respectively sink, is also called an *input(-node)*, respectively an *output(-node)* of the graph.

Path, circuit, loop. A path ρ is a sequence of nodes $i_1, i_2, \ldots, i_p$, $p > 1$, such that $i_j \in \pi(i_{j+1}), j = 1, \ldots, p-1$. Node i_1 is the initial node and i_p is the final one of this path. The *length* of the path is equal to the sum of the lengths of the arcs of which it is composed, the lengths of the arcs being 1. The length of path ρ is denoted $|\rho|_l$ (equal to $p-1$ in the above example). The subscript 'l' here refers to the word 'length' (later on another subscript 'w' will appear for a different concept). Equivalently, one also says that a path is a sequence of arcs which connects a sequence of nodes. An *elementary path* is a path in which no node appears more than once. A circuit is a path where the initial and the final node coincide. An *elementary circuit* $i_1, i_2, \ldots, i_p = i_1$ is a circuit in which the path $i_1, i_2, \ldots, i_{p-1}$ is elementary. A loop is a circuit involving a single node. A digraph is said to be *acyclic* if it contains no circuits.

Descendant, ascendant. The set of descendants $\sigma^+(i)$ of node i consists of all nodes j such that a path exists from i to j. Similarly the set of ascendants $\pi^+(i)$ of node i is the set of all nodes j such that a path exists from j to i. One has, e.g., $\pi^+(i) = \pi(i) \cup \pi(\pi(i)) \cup \ldots$. The

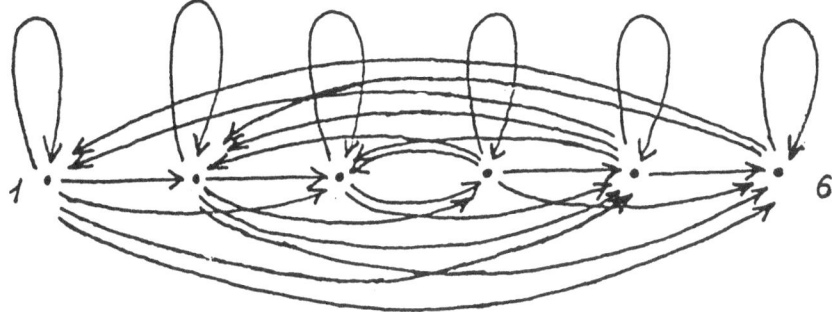

Figure 6: Precedence graph of the matrix M

mapping $i \mapsto \pi^*(i) = \{i\} \cup \pi^+(i)$ is the transitive closure of π; the mapping $i \mapsto \sigma^*(i) = \{i\} \cup \sigma^+(i)$ is the transitive closure of σ.

Strongly connected graph. A graph is called strongly connected if for any two different nodes i and j there exists a path from i to j. Equivalently, $i \in \sigma^*(j)$ for all $i, j \in \mathcal{V}$, with $i \neq j$. Note that according to this definition an isolated node, with or without a loop, is a strongly connected graph.

Consider a graph $\mathcal{G} = (\mathcal{V}, \mathcal{E})$. If we associate a real number a_{ij} to each arc $(j, i) \in \mathcal{E}$, then $\mathcal{G}$ is called a *weighted graph*. The quantity a_{ij} is called the *weight* of arc (j, i). Note that the second subscript of a_{ij} refers to the initial (and not the final) node. The reason is that in the algebraic context we work with column vectors (and not with row vectors).

In the following definition the starting point is a square matrix, the entries of which may again assume the 'value' ε.

Definition .12 (Precedence graph) *The precedence graph of a square matrix A, of size $n \times n$, is a weighted digraph with n nodes and an arc (j, i) if $a_{ij} \neq \varepsilon$, in which case the weight of this arc receives the numerical value of a_{ij}. The precedence graph is denoted $\mathcal{G}(A)$.*

It is not difficult to see that any weighted digraph $\mathcal{G} = (\mathcal{V}, \mathcal{E})$ is the precedence graph of an appropriately defined square matrix. The weight a_{ij} of the arc from node j to node i is defined as the ij-th entry of a matrix A. If an arc does not exist, the corresponding entry of A becomes ε. The matrix A thus defined has $\mathcal{G}$ as its precedence graph. As an example, the precedence graph of (26) is given in Figure 6; it has many circuits. The precedence graph of A from (21), not shown in a figure here, is acyclic. This latter property yields $A^k = \varepsilon$ for $k \geq n$.

Let $\mathcal{G} = (\mathcal{V}, \mathcal{E})$ be a weighted digraph with n nodes. The weights are combined such as to form the $n \times n$ matrix A. The numerical value of a_{ij}

equals the weight of the arc from node j to node i. If no such arc exists, then $a_{ij} = \varepsilon$. As seen before, the element (i,j) of $A^k = A \otimes \cdots \otimes A$, considered within the max-plus algebra denotes the maximum weight with respect to all paths of length k which go from node j to node i. If no such path exists, then $(A^k)_{ij} = \varepsilon$. Within this algebra, ε gets assigned the numerical value $-\infty$. The weight of a path ρ is denoted $|\rho|_w$.

Definition .13 *The mean weight of a path is defined as the sum of the weights of the individual arcs of this path, divided by the length of this path. If the path is denoted by ρ, then the mean weight equals $|\rho|_w/|\rho|_l$. If such a path is a circuit one talks about the mean weight of the circuit, or simply the cycle mean.*

We are interested in the maximum of these cycle means, where the maximum is taken over all circuits in the graph. This number will be called the *maximum cycle mean*.

4.2 Spectral Theory of Matrices

Given a matrix A with entries in the max-plus algebra, we consider the problem of existence of eigenvalues and eigenvectors, that is, the existence of λ and v such that:

$$Av = \lambda v \ . \tag{27}$$

For example,

$$\begin{pmatrix} 3 & 7 \\ 2 & 4 \end{pmatrix} \begin{pmatrix} 2.5 \\ e \end{pmatrix} = 4.5 \begin{pmatrix} 2.5 \\ e \end{pmatrix} \ .$$

Thus it is seen that the matrix A of (4) has an eigenvalue 4.5. To exclude degenerate cases, it is assumed that not all elements of v are identical to ε. Equation (7) is also valid in the current setting. If x_0 is an eigenvector of A, with corresponding eigenvalue λ, then the solution of the difference equation (11) can be written as

$$x(k) = \lambda^k x_0 \quad (= \lambda^k \otimes x_0), \quad k = 0, 1, \ldots \tag{28}$$

The numerical evaluation of λ^k in this formula equals $k\lambda$ in conventional analysis. The eigenvalue λ can be interpreted as the cycle time (defined as the inverse of the throughput) of the underlying system; each node of the corresponding network becomes active every λ units of time, as it follows straightforwardly from (28). Also, the relative order in which the nodes become active for the k-th time, as expressed by the components $x_i(k)$, is exactly the same as the relative order in which the nodes become active for the $(k + 1)$-st time. More precisely, equation (28) yields $x_l(k + 1) - x_j(k + 1) = x_l(k) - x_j(k)$, $j, l = 1, \ldots, n$. Thus the solution (28) exhibits a kind of periodicity. Procedures exist for the calculation of eigenvalues and eigenvectors; an efficient one is the procedure known as Karp's algorithm [5].

The main result on eigenvalues in the max-plus algebra is as follows.

Theorem .14 *We are given a square matrix A. If $\mathcal{G}(A)$ is strongly connected, there exists one and only one eigenvalue and at least one eigenvector. The eigenvalue is equal to the maximum cycle mean of the graph:*

$$\lambda = \max_{\zeta} \frac{|\zeta|_{\mathrm{w}}}{|\zeta|_{\mathrm{l}}} ,$$

where ζ ranges over the set of circuits of $\mathcal{G}(A)$.

Proof

Existence of x and λ. Define the matrix B by subtracting λ, in the conventional way, from each entry of A, where $\lambda = \max_{\zeta} |\zeta|_{\mathrm{w}}/|\zeta|_{\mathrm{l}}$. Now the maximum circuit weight of $\mathcal{G}(B)$ is e. Hence B^* and $B^+ = BB^*$ exist. The matrix B^+ has some columns with entries e on the diagonal. Indeed, we can pick a node k on a circuit $\xi \in \arg\max_{\zeta} |\zeta|_{\mathrm{w}}/|\zeta|_{\mathrm{l}}$. The maximum weight of all paths from k to k is e. Therefore we have $e = B^+_{kk}$. Then:

$$B^+_{\cdot k} = B^*_{\cdot k} \Rightarrow BB^*_{\cdot k} = B^+_{\cdot k} = B^*_{\cdot k} \Rightarrow AB^+_{\cdot k} = \lambda B^+_{\cdot k} ,$$

where $B_{\cdot k}$ denotes column k of B. Hence $x = B^+_{\cdot k}$ is an eigenvector. The set of nodes of $\mathcal{G}(A)$ corresponding to indices of the nonzero components of x is called the *support* of x.

If $\mathcal{G}(A)$ is strongly connected, the support of x contains all nodes. Let us suppose that the support of x does not cover the whole graph. Then there are arcs going from the support of x to other nodes because the graph $\mathcal{G}(A)$ is strongly connected. Then the support of Ax would be larger than the support of x which is contradicted by $Ax = \lambda x$.

Uniqueness of λ. If λ satisfies Equation (27), we have $(Ax)_1 = \lambda x_1$ and there exists a component x_{i_1} such that $A_{1i_1} x_{i_1} = \lambda x_1$. Then $(Ax)_{i_2} =$

λx_{i_1} and there exists a component x_{i_2} such that $A_{i_1 i_2} x_{i_2} = \lambda x_{i_1}$ and so on until we reach a component x_{i_l} that we have already met. In this way we have defined a circuit $\beta = (i_l, i_m, \ldots, i_{l+1}, i_l)$ such that:

$$A_{i_l i_{l+1}} A_{i_{l+1} i_{l+2}} \cdots A_{i_m i_l} x_{i_{l+1}} x_{i_{l+2}} \cdots x_{i_m} x_{i_l} = \lambda^{m-l+1} x_{i_l} x_{i_{l+1}} \cdots x_{i_m} .$$

Therefore, because $x_k \neq \varepsilon$ for all k, λ^{m-l+1} is the weight of circuit of length $m - l + 1$. Hence λ is the average weight of circuit β.

Let us now take any circuit $\gamma = (i_1, \ldots, i_p, i_1)$ such that its nodes belong to the support of x (here any node of $\mathcal{G}(A)$). We have:

$$A_{i_2 i_1} x_{i_1} \leq \lambda x_{i_2} ,$$

$$\vdots$$

$$A_{i_p i_{p-1}} x_{i_{p-1}} \leq \lambda x_{i_p} ,$$

$$A_{i_1 i_p} x_{i_p} \leq \lambda x_{i_1} .$$

Hence, by $\otimes$-multiplying these inequalities and because $x_k \neq \varepsilon$ for all k, one sees that λ is greater than the average weight of γ. Therefore λ is the maximum cycle mean and it is unique.

∎

Remark .15 It is important to understand the role of the support of x in the previous proof. If $\mathcal{G}(A)$ is not strongly connected, the support of x is not necessarily the whole set of nodes and, in general, there is not a unique eigenvalue.

∎

Example .16 With the assumption of Theorem .14 , the uniqueness of the eigenvector is not assured as is shown by

$$\begin{pmatrix} 1 & e \\ e & 1 \end{pmatrix} \begin{pmatrix} e \\ -1 \end{pmatrix} = \begin{pmatrix} 1 \\ e \end{pmatrix} = 1 \begin{pmatrix} e \\ -1 \end{pmatrix},$$

and

$$\begin{pmatrix} 1 & e \\ e & 1 \end{pmatrix} \begin{pmatrix} -1 \\ e \end{pmatrix} = \begin{pmatrix} e \\ 1 \end{pmatrix} = 1 \begin{pmatrix} -1 \\ e \end{pmatrix}.$$

∎

Example .17 The following example is a trivial counterexample to the uniqueness of the eigenvalue if the graph is not strongly connected:

$$\begin{pmatrix} 1 & \varepsilon \\ \varepsilon & 2 \end{pmatrix} \begin{pmatrix} e \\ \varepsilon \end{pmatrix} = 1 \begin{pmatrix} e \\ \varepsilon \end{pmatrix}, \quad \begin{pmatrix} 1 & \varepsilon \\ \varepsilon & 2 \end{pmatrix} \begin{pmatrix} \varepsilon \\ e \end{pmatrix} = 2 \begin{pmatrix} \varepsilon \\ e \end{pmatrix}.$$

∎

Example .18 The matrix M of (26) has a unique eigenvalue:

$$\begin{pmatrix} 6 & \varepsilon & \varepsilon & \varepsilon & 6 & 5 \\ 9 & 8 & \varepsilon & 8 & 9 & 8 \\ 6 & 10 & 7 & 10 & 6 & \varepsilon \\ \varepsilon & 7 & 4 & 7 & \varepsilon & \varepsilon \\ 6 & 10 & 7 & 10 & 6 & \varepsilon \\ 9 & 8 & \varepsilon & 8 & 9 & 8 \end{pmatrix} \begin{pmatrix} e \\ 3 \\ 3.5 \\ .5 \\ 3.5 \\ 3 \end{pmatrix} = 9.5 \begin{pmatrix} e \\ 3 \\ 3.5 \\ .5 \\ 3.5 \\ 3 \end{pmatrix}.$$

It follows that the eigenvalue equals 9.5, which means in more practical terms that the manufacturing system 'delivers' an item (a product or a machine) at all of its output channels every 9.5 units of time. ∎

The eigenvector of this latter example is also unique, apart from adding the same constant to all components. If v is an eigenvector, then cv, where c is a scalar, is also an eigenvector, as it follows directly from the definition of eigenvalue. It is possible that several eigenvectors can be associated with the only eigenvalue of a matrix, i.e. eigenvectors may not be identical up to an additional constant as shown in Example .16.

5 A Stochastic Extension

The evolution equation studied in this section is

$$x(k+1) = A(k) \otimes x(k), \quad k = 0, 1, 2, \ldots, \tag{29}$$

with some initial condition $x(0)$. Some (or all) entries of $A(k)$ are stochastic. We assume that

- the underlying distribution functions do not depend on k.

- these stochastic entries can assume only a finite number of different values. It will also be assumed that these values are finite, though the method to be described can be generalized to the case that $-\infty$ is also allowed as a value.

- $A(k)$ and $A(l)$ are independent stochastic matrices for $k \neq l$. Problems where $A(k)$ and $A(k+1)$ are correlated can be treated also, provided there exists a model with the Markov property that describes the evolution of $A(k)$, $k = 0, 1, \ldots$. In such a case the latter model would be added to (29) and the theory to be described should be applied to this augmented model. For reason of simplicity, we will not explicitly deal with such problems.

- no correlation between stochastic entries of $A(k)$ exists, though such correlations can be treated rather routinely.

- $\mathcal{G}(A(k))$ is strongly connected. (If this assumption is true for one k, it automatically is true for all k due to the second assumption above.)

The quantity of central interest in this section is

$$\lim_{k \to \infty} E(x_i(k+1) - x_i(k)), \tag{30}$$

for an arbitrary i, being the average cycle time for component i. This quantity is a kind of 'average cycle time'; it can been proved [1] that this average cycle time is independent of i. The method of calculation of the average cycle time will be shown by means of a simple example. Consider the case that $x \in R^2$ and that for each k the matrix A is one of the following two matrices

$$\begin{pmatrix} 3 & 7 \\ 2 & 4 \end{pmatrix}, \begin{pmatrix} 3 & 5 \\ 2 & 4 \end{pmatrix}.$$

Both matrices occur with probability $1/2$ and there is no correlation in time. Starting from an arbitrary $x(0)$-vector, say $x(0) = (0,2)'$, we will set up the reachability tree of all possible states x. This is indicated the following table, being a table of transitions. In order to get a concise notation, the different state vectors are indicated by n_i, $i = 1, \ldots$. The table has been obtained in the following way. The starting point is $n_1 \stackrel{\text{def}}{=} (0,2)'$. From there, two states can be reached in one step: $(9,6)'$ or $(7,6)'$, depending on which A-matrix occurs. The states will be normalized such that the first component equals zero. This results in $(0,-3)'$ and $(0,-1)'$. (Other normalizations are

Table 2: Transitions of stochastic states

initial state	$a_{12} = 7$	$a_{12} = 5$
$n_1 = (0,2)'$	$n_2 + 91$	$n_3 + 71$
$n_2 = (0,-3)'$	$n_4 + 41$	$n_3 + 31$
$n_3 = (0,-1)'$	$n_2 + 61$	$n_3 + 41$
$n_4 = (0,-2)'$	$n_2 + 51$	$n_3 + 31$

possible, and they will lead to the same results.) Both states are new and are therefore added to the list, as n_2 and n_3 respectively. Now we take n_2 as the starting point. Two states can be reached from there: $(4,2)'$ and $(3,2)'$, or, after normalization, $(0,-2)'$ and $(0,-1)'$. Only the first of these states is new and will be added to the list as the next state n_4. In this way we continue: from all states obtained sofar we construct the states which can be reached from there in one step. If a state is found which did not exist sofar, it is added to the list. For the current example it turns out that there exist four different states. The notation $n_i + j1$ in the table refers to the state n_i of which all components are increased by the number j. One directly notices, by viewing the table, that the system never returns to n_1. Hence this node is a transient one. In the stationary situation a Markov chain results with the three states n_2, n_3 and n_4. Let us be slightly more explicit. The elements of this Markov chain, to be denoted by $z(k)$, are, by construction,

$$z(k) = \begin{pmatrix} 0 \\ x_2(k) - x_1(k) \end{pmatrix}.$$

It is easily shown that

$$z(k+1) = \begin{pmatrix} 0 \\ (Ax(k))_2 - (Ax(k))_1 \end{pmatrix} = \begin{pmatrix} 0 \\ (Az(k))_2 - (Az(k))_1 \end{pmatrix},$$

and hence the process $\{z(k)\}$ is indeed Markovian. The transition matrix of the Markov chain is

$$\begin{pmatrix} 0 & 1/2 & 1/2 \\ 1/2 & 1/2 & 1/2 \\ 1/2 & 0 & 0 \end{pmatrix}.$$

The stationary distribution of this chain is easily calculated to be

$$\Pr(n_2) = 1/3, \ \Pr(n_3) = 1/2, \ \Pr(n_4) = 1/6.$$

The average cycle time becomes

$$\Pr(n_2)(4\Pr(A_1) + 3\Pr(A_2)) + \Pr(n_3)(6\Pr(A_1) + 4\Pr(A_2)) \\ + \Pr(n_4)(5\Pr(A_1) + 3\Pr(A_2)) = 13/3,$$

where the coefficients are the appropriate numbers out of the table above. The first term in this expression for instance, $\Pr(n_2)(4\Pr(A_1 + 3\Pr(A_2))$, is obtained as follows. If the state is in n_2, then this happens with (stationary) probability $\Pr(n_2)$. The next step either leads to n_4, with probability $\Pr(A_1)$ and obtained after 4 time units (see Table 2), or it leads to n_3, with probability $\Pr(A_2)$ and obtained after 3 time units. The other terms are obtained similarly. It is the quantity at the right-hand side,13/3, which equals the expression in (30).

This example described a method to calculate the average cycle time. The crucial feature in this method is that the number of different normalized state vectors is finite.

6   Counter versus Dater Description

The new concepts of counter and dater descriptions will be explained by means of the equation

$$x(k + 1) = M \otimes x(k), \tag{31}$$

where M equals the matrix introduced in (26). The variable $x_i(k)$ denotes the time instant at which output i delivers its k-th item (being a product or a machine). We now introduce a quantity $\chi_i(t)$ which is related to $x_i(k)$. The argument t of $\chi_i(t)$ refers to the actual clocktime and $\chi_i(t)$ itself refers to the number of times that output i has delivered an item up to (and including) time t. The quantity χ_i can henceforth only assume the values $0, 1, 2, \ldots$. Considering the numerical values of the entries of M, it easily follows that

$$\chi_1(t) = \min\left(\chi_1(t-6) + 1, \chi_5(t-6) + 1, \chi_6(t-5) + 1\right) .$$

For χ_2 one similarly obtains

$$\chi_2(t) = \min(\chi_1(t-9) + 1, \chi_2(t-8) + 1, \chi_4(t-8) + 1, \\ \chi_5(t-9) + 1, \chi_6(t-8) + 1),$$

etc. One can compactly write

$$\chi(t) = \overline{A}_1 \otimes \chi(t-1) \oplus \overline{A}_2 \otimes \chi(t-2) \oplus \cdots \oplus \overline{A}_l \otimes \chi(t-l) \qquad (32)$$

for some finite l ($l = 10$ in the example with the M-matrix). This latter equation must be read in the so-called min-plus algebra. This min-plus algebra equals the max-plus algebra, except that everywhere where the max-operation appears in the max algebra, one now must read the min-operation. Equation (32) is a higher order difference equation in the min-plus algebra. It can be transformed to a first order differential equation similar to the way as explained in Section 2.6 for scalar higher order equations in the max-plus algebra. Equation (31) (in the max-plus algebra) and Equation (32) (in the min-plus algebra) describe the same system. Equation (32) is referred to as the *counter description* and the other one as the *dater description*. The word 'dater' must be understood as 'timer', but since the word 'time' and its declinations are already used in various ways, the word 'dater' is used. The awareness of these two different descriptions for the same problem has far reaching consequences for the theory of discrete event systems.

The reader should contemplate that the stochastic problem (in which some of the a_{ij} are stochastic) is not very suitable to be given in the counter description, since then the delays in (32) would be stochastic.

7 The z-Transform

Conventional linear systems with inputs and outputs are of the form (13), though (13) itself has the max-plus algebra interpretation. This equation, now considered in the conventional way, is a representation of a linear system in the time domain. Its representation in the z-domain equals

$$Y(z) = C(zI - A)^{-1} BU(z) \ ,$$

where $Y(z), U(z)$ are defined by

$$Y(z) = \sum_{i=0}^{\infty} y(i) z^{-i}, \quad U(z) = \sum_{i=0}^{\infty} u(i) z^{-i} \ ,$$

where it is tacitly assumed that the system was at rest for $t \leq 0$ and where I refers to the unit matrix in the conventional algebra. The matrix $H(z) \stackrel{\text{def}}{=} C(zI - A)^{-1} B$ is called the transfer matrix of the system. The notion of transfer matrix is especially useful when subsystems are combined to build larger sytems, by means of parallel, series and feedback connections, see [6].

In the max-plus algebra context, the z-transform also exists, but here it is customary to refer to it as the γ-transform where γ operates as z^{-1}. For instance, the γ-transform of u is defined as

$$U(\gamma) = \bigoplus_{i=0}^{\infty} u(i) \otimes \gamma^i \;,$$

and $Y(\gamma)$ and $X(\gamma)$ are defined likewise. Multiplication of (14) by γ^k yields

$$\left.\begin{array}{rcl} \gamma^{-1} x(k+1)\gamma^{k+1} & = & A \otimes x(k)\gamma^k \oplus B \otimes u(k)\gamma^k, \\ y(k)\gamma^k & = & C \otimes x(k)\gamma^k \;. \end{array}\right\} \qquad (33)$$

If these equations are summed with respect to $k = 0,\ldots,$ and if we add $\gamma^{-1}x_0$ to both sides then we obtain

$$\left.\begin{array}{rcl} \gamma^{-1} X(\gamma) & = & A \otimes X(\gamma) \oplus B \otimes U(\gamma) \oplus \gamma^{-1}x_0 \;, \\ Y(\gamma) & = & C \otimes X(\gamma) \;. \end{array}\right\} \qquad (34)$$

The first of these equations can be solved by first multiplying (max-plus algebra), equivalently adding (conventional), left- and right-hand side by γ and then repeatedly substituting the right-hand side for $X(\gamma)$ within this right-hand side. This results in

$$X(\gamma) = (\gamma A)^*(\gamma B U(\gamma) \oplus x_0) \;.$$

Thus we obtain $Y(\gamma) = H(\gamma)U(\gamma)$, provided that $x_0 = \varepsilon$, and where the transfer matrix $H(\gamma)$ is defined by

$$H(\gamma) = C \otimes (\gamma A)^* \otimes \gamma \otimes B = \gamma CB \oplus \gamma^2 CAB \oplus \gamma^3 CA^2 B \oplus \cdots \qquad (35)$$

The expression $Y(\gamma) = H(\gamma)U(\gamma)$ is the max-plus algebra equivalent of $Y(z) = H(z)U(z)$ in the conventional system theory. It can also conveniently be used for building larger systems from subsystems. The transfer matrix is defined by means of an infinite series and the convergence depends on the value of γ. If the series is convergent for $\gamma = \gamma'$, then it is also convergent for all γ's which are smaller than γ'. If the series does not converge, it still has a meaning as a formal series.

Exactly as in conventional system theory, the product of two transfer matrices (of which it is tacitly assumed that the sizes of these matrices is such that the multiplication is possible), is a new transfer matrix which refers to a system which consists of the original systems put in a series connection. In the same way, the sum of two transfer matrices refers to two systems put in parallel. This section will be concluded by an example of such a parallel connection.

We are given two systems. The first one is given in (15), and is characterized by the 1×1 transfer matrix

$$H_1 = \varepsilon\gamma \oplus 11\gamma^2 \oplus 14\gamma^3 \oplus 20\gamma^4 \oplus 24\gamma^5 \oplus 29\gamma^6 \oplus \cdots$$

It is easily shown that this series converges for $\gamma \leq 4.5$; this bound on γ corresponds to the eigenvalue of A. The second system is given by

$$x(k+1) = \begin{pmatrix} e & \varepsilon & 4 \\ 1 & 1 & \varepsilon \\ \varepsilon & 6 & 3 \end{pmatrix} x(k) \oplus \begin{pmatrix} \varepsilon \\ 2 \\ e \end{pmatrix} u(k) ,$$

$$y(k) = (\, 1 \quad 1 \quad 4 \,)x(k) ,$$

and its transfer matrix is

$$H_2 = 4\gamma \oplus 12\gamma^2 \oplus 15\gamma^3 \oplus 18\gamma^4 \oplus 23\gamma^5 \oplus 26\gamma^6 \oplus \cdots$$

The transfer matrix of the two systems put in parallel has size 1×1 again (one can talk about a transfer function) and is obtained as

$$H_{\text{par}} = H_1 \oplus H_2 = 4\gamma \oplus 12\gamma^2 \oplus 15\gamma^3 \oplus 20\gamma^4 \oplus 24\gamma^5 \oplus 29\gamma^6 \oplus \cdots \quad (36)$$

A transfer function can easily be visualized. If $H(\gamma)$ is a scalar function, i.e. the system has one input and one output, then it is a continuous and piecewise linear function. As an example, the transfer function of the parallel connection considered above is pictured in Figure 7.

Above it was shown how to derive the transfer matrix of a system if the representation of the system in the 'time domain' is given. This time domain representation is characterized by the matrices A, B and C. Now one could pose the opposite question; how to obtain a time domain representation, or equivalently, how to find A, B and C if the transfer matrix is given. A partial answer to this question is given in [7]. For the example above, one would like to obtain a time domain representation of the two systems put in parallel starting from (36). This avenue will not be pursued here.

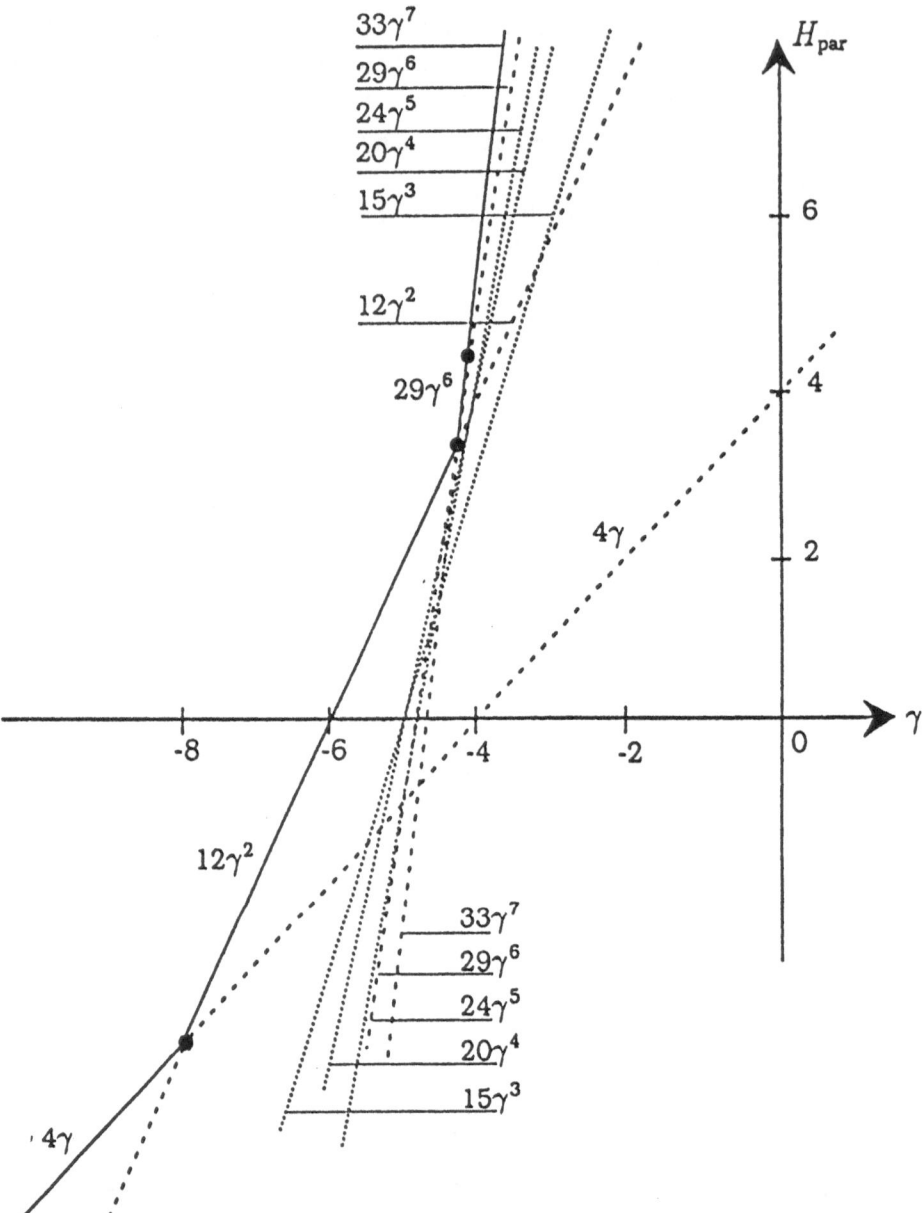

Figure 7: The transfer function H_{par} as a function of γ

8 Conclusions

This paper has shown some recent and exciting developments in the theory of discrete event dynamic systems (DEDS). The theory came into existence around the beginning of the years eighty, though the max-plus algebra was studied earlier, see [3]. The theory of DEDS is based on some other disciplines in (applied) mathematics, specifically on linear systems theory (see [6]) and graph theory (see [4]). The theory of Petri nets is a related topic for which [8] is a good introduction. A comprehensive treatment of DEDS can be found in [1]. The current paper has been based on Chapter 1 of this book for an essential part. In this reference, as in [3], the underlying algebra is not limited to the max-plus algebra, but rather an axiomatic and more abstract point of view is given. It is believed that the development of the theory is only in its childhood and that many more results will follow. The applications of the theory to practical problems is partly parallel to the applications of timed Petri nets. The applications specifically lie in the areas of parallel and distributed processing, where synchronization plays a role. An application to time-table dependent transportation networks is given in [2].

References

[1] F. Baccelli, G. Cohen, G.J. Olsder, and J.P.Quadrat. *Synchronization and Linearity*. John Wiley, 1992.

[2] J.G. Braker. Max-algebra modelling and analysis of time-table dependent networks. In *Proceedings of the first European Control Conference*, pages 1831–1836. Hermes, Paris, 1991.

[3] R.A. Cuninghame Green. *Minimax Algebra*. Lecture Notes in Economics and Mathematical Systems, no 166. Springer Verlag, 1979.

[4] M. Gondran and M. Minoux. *Graphs and Algorithms*. John Wiley, 1986.

[5] Richard M. Karp. A characterization of the minimum cycle mean in a digraph. *Discrete Mathematics*, 23:309–311, 1978.

[6] Huibert Kwakernaak and Raphael Sivan. *Linear Optimal Control Systems*. Wiley-Interscience, New York, 1972.

[7] G.J. Olsder and R.E. de Vries. On an analogy of minimal realizations in conventional and discrete- event dynamic systems. In P.Varaiya and A.B. Kurzhanski, editors, *Discrete Event Systems: Models and Applications*, volume 103 of *Lecture Notes in Control and Information Sciences*, pages 149–161. Springer Verlag, Berlin, 1988.

[8] James L. Peterson. *Petri net theory and the modeling of systems*. Prentice Hall, Englewood Cliffs, N.J. 07632, 1981.

Acoustical detection of obstructions in a pipe with a temperature gradient

S.W. Rienstra

Institute for Mathematics Consulting
Faculty of Mathematics and Computing Science
Eindhoven University of Technology
P.O. Box 513, 5600 MB Eindhoven, The Netherlands

Abstract

Exhaust pipes of furnaces and similar equipment (e.g. for gasification) may get clogged up by deposition of sticky particles (ashes, tar, soot) carried by the gas mixture. The possibility of detecting the presence of such an obstruction by the reflection of an acoustic wave is investigated by a mathematical model. An important part of the problem is the considerable temperature variation along the pipe. It is argued that the high frequency components of the wave may be not reliable because of refraction and spurious reflection effects caused by this temperature gradient. Therefore, the reflection of low-frequency (one mode propagating) waves is investigated. It appears that the reflection becomes significant for a blockage area of more than 50%.

1. INTRODUCTION

Mathematics has, historically, its sources of inspiration in applications [1]. Indeed, unexpected questions from practice force one to go off the beaten track. Also it is easier to portray an abstraction with a concrete example at hand. Therefore, most mathematics is applied, or at least applicable or emerging from applications. And if we give up the mathematico-centric point of view there is even more applied mathematics hidden in the various theoretical disciplines of science.

Most of this applied mathematics is now applied to a practical problem as existing mathematical results, rather than mathematics invented for the problem. Mathematics is fed back to the application. However, there is more.

A. van der Burgh and J. Simonis (eds.), Topics in Engineering Mathematics, 151–179.
© 1992 Kluwer Academic Publishers.

Before mathematics may be applied, the problem must be mathematized: formulated in equations and formula's, to render it amenable to formal manipulations. This is called mathematical modelling. If the problem considered does not have an accepted model, the modelling is evidently very important, and the mathematical discipline with this crucial rôle of modelling is (now) called industrial mathematics [2].

If the problem has a longer history (like from fluid mechanics, acoustics, elasticity, etc.) there is usually a universally accepted general model. Such a model is, however, often too general to be useful and it pays to break it down a bit to a simplified model which contains exactly the essential elements of the problem. In that case mathematics serves also as a language in which we can express precisely the various elements of the model and formulate exactly what we mean with "essential elements".

Finally the main rôle of mathematics comes into play, as the model may be analysed by mathematical techniques, not rarely originating from other scientific disciplines or problems, and thus realizing a fruitful cross-fertilization [3].

This last type of mathematics finds its area of applications mainly in the traditional engineering problems, which is why we call it *Engineering Mathematics*: applied mathematics where the problem is central and the mathematics most important.

In the present paper we present a problem of this type. It is typically engineering: acoustic wave propagating and scattering in an exhaust pipe with temperature gradients. The mathematics is, although not very difficult, not trivial. Furthermore, advantageous for the present introductory purposes, the problem consists of a number of subproblems, each requiring a suitable mathematical technique that heavily leans on the physics of the problem.

2. THE PROBLEM

The exhaust ducting of furnaces and gasification installations transfers a mixture of sometimes highly pressurized gases from the high temperature furnace chamber to an area of lower temperatures. Sticky particles of ashes, tar, and soot are carried away with this gas mixture.

During the design of a new type of industrial gasification furnace doubts were raised as to what extent the liquid particles would be deposited on the (colder) wall. If this really occurs, in spite of the precautions taken, it is necessary to measure it during the process in operation, because the contamination will accumulate at about the same location in the pipe and the resulting constriction may finally block the pipe.

Due to the very polluting and hostile conditions inside the pipe it is very difficult to apply measuring devices directly. A possible alternative proposed is to measure the contamination acoustically. At a distance sufficiently downstream in the pipe a transducer mounted in the wall generates an acoustic wave, which is reflected by the possible obstacle in the pipe. By measuring the reflected wave this obstacle may be detected.

The present paper reports a theoretical investigation to quantify this idea by a mathematical model, and to find out under which conditions a sufficiently discriminative reflection is obtained, not afflicted or spoiled by unwanted other reflections. In particular, the reflection from the pipe-furnace connection, and possible reflections caused by the temperature gradients may mask the reflections from the sought constriction.

To distinguish between incident and reflected wave easily it is probably most convenient to deal with a wave of finite duration. However, if the wave is very much pulse-like, the incident wave form is practically unknown, and the spectrum contains many high frequency components which propagate down the duct mainly via spiralling paths (modes), in contrast to the straight, axial path of the low frequency components [4]. Apart from the longer travel time, these spiralling modes are also very vulnerable to the temperature gradients. If the pitch of the spiral is small enough, a temperature increase will cause a reflection of the sound at a point in the duct determined only by an unfortunate but otherwise fortuitous combination of problem parameters. A reflection which has nothing to do with any obstacle in the duct!

For this reason it may be safer to restrict ourselves to the low frequencies. This, however, has the disadvantage that the open end generates a strong reflection and at the same time any obstacle a weak reflection, and that the front end of the reflected wave is not well defined. On the other hand, the spatial distribution of the wave is very well known, and measuring the pressure at one point is sufficient to know the wave.

The model we will elaborate will contain as essential elements:

(i)   temperature variation, both axially and radially, in the duct to include the wave form variations attended with it, and to predict possible spurious reflections;

(ii)   reflection and scattering by an annular constriction;

(iii)   reflection by the flanged open end connecting the pipe to the large furnace chamber.

This rather ambitious object to incorporate the extra complexities of a non-uniform medium and reflection effects needs, however, to be accompanied by limitations in other respects to keep sight of the physically essential processes. Only after that we understand the problem well we are ready to turn to a more complete model, using a numerical approach. Therefore, the problem will be modelled to allow an analytical treatment as much as is reasonable.

3. THE MODEL

3.1. Differential equations

As a fluid mechanical phenomenon, sound has to satisfy the conservation laws of fluid motion. Introduction of the usual acoustical approximations then will yield a version of the wave equation pertaining to the present problem. For a simple and uniform, quiescent medium the equation for acoustic perturbations is just the standard wave equation, cited in many textbooks [4,5]. For the present configuration, however, with a varying temperature the result is different and not entirely straightforward. It is therefore instructive to derive the equations in detail and see which assumptions are underlying the model.

In tensor notation we have [4,5,6]:
the equation of mass conservation

$$\frac{\partial}{\partial t}\rho + \nabla \cdot (\rho \mathbf{v}) = 0 \ , \tag{3.1}$$

the equation of momentum conservation

$$\frac{\partial}{\partial t}(\rho \mathbf{v}) + \nabla \cdot (-\mathbf{P} + \rho \mathbf{v} \mathbf{v}) = \mathbf{f} \ , \tag{3.2}$$

and the equation of energy conservation

$$\frac{\partial}{\partial t}(\rho \epsilon + \tfrac{1}{2}\rho v^2) + \nabla \cdot (\rho \epsilon \mathbf{v} + \tfrac{1}{2}\rho v^2 \mathbf{v}) =$$

$$\nabla \cdot (\mathbf{v} \cdot \mathbf{P}) - \nabla \cdot \mathbf{q} + \mathbf{f} \cdot \mathbf{v} \ , \tag{3.3}$$

where t is the time, ρ is the density, $\mathbf{v}$ is the velocity, $\mathbf{P}$ is the stress tensor, $\mathbf{f}$ is an external force (like gravity), ϵ is the internal energy, and $\mathbf{q}$ is the heat flux due to heat conduction. The stress tensor is split up into a normal stress and a shear stress component

$$\mathbf{P} = -p\,\mathbf{I} + \boldsymbol{\tau} \tag{3.4}$$

where p is the hydrostatic pressure, $\boldsymbol{\tau}$ the viscous stress tensor and $\mathbf{I}$ denotes the identity tensor with elements δ_{ij}.

The second law of thermodynamics for reversible processes relates for a fluid element the internal energy change $d\epsilon$, thermodynamic pressure p_{th} and volume change $d\rho^{-1}$ to temperature T and entropy change ds

$$T\,ds = d\epsilon + p_{th}\,d\rho^{-1} \ . \tag{3.5}$$

Stokes' hypothesis is that the fluid is locally in thermodynamic equilibrium so that thermodynamic pressure and hydrostatic pressure are equivalent $(p_{th} = p)$, and p, ρ and s are related via a single constitutive equation of state

$$p = p(\rho, s) \ . \tag{3.6}$$

If we neglect for the acoustic perturbations viscous dissipation ($\sim \tau$) and heat transfer ($\sim$ q) then equations (3.2) and (3.3) may be simplified. Equation (3.3) becomes

$$\rho \frac{D}{Dt}\epsilon = -p\nabla \cdot \mathbf{v} \tag{3.7}$$

with convective derivative

$$\frac{D}{Dt} = \frac{\partial}{\partial t} + \mathbf{v} \cdot \nabla \ . \tag{3.8}$$

It is important to note that since the convective derivative just describes temporal variations of a fluid element travelling with the fluid flow, equation (3.5) is only to be interpreted as

$$T\frac{D}{Dt}s = \frac{D}{Dt}\epsilon + p\frac{D}{Dt}\rho^{-1} \ , \tag{3.9}$$

which reduces, with (3.7) and (3.1), to the equation of isentropy along a streamline, $Ds/Dt = 0$. All in all, we have now

$$\frac{D}{Dt}\rho + \rho\nabla \cdot \mathbf{v} = 0 \tag{3.10}$$

$$\rho\frac{D}{Dt}\mathbf{v} + \nabla p = \mathbf{f} \tag{3.11}$$

$$\frac{D}{Dt}s = 0 \ . \tag{3.12}$$

Equation (3.6) with (3.12) can be written out as

$$\frac{D}{Dt}p = c^2\frac{D}{Dt}\rho \tag{3.13}$$

where $c^2 = (\partial p/\partial \rho)_s$ is the square of the local sound speed. For an ideal gas is

$$p = \rho R T \qquad (3.14)$$

(R the gas constant) and ϵ only dependent on T, so that the specific heat capacities (for constant volume and pressure) are given by

$$c_v = d\epsilon/dT , \quad c_p = R + c_v$$

while

$$ds = c_v \frac{dp}{p} + c_p \frac{d\rho^{-1}}{\rho^{-1}} .$$

As a result is

$$c^2 = \gamma p/\rho = \gamma R T \qquad (3.15)$$

where $\gamma = c_p/c_v$. So in general c is a function of the temperature alone. In the present problem experiments have shown that the medium behaves like an ideal gas with constant specific heat capacities. This implies that c^2 varies linearly with T.

If we write the variables as a mean stationary component plus an acoustic perturbation

$$p = p_0 + p' , \quad \rho = \rho_0 + \rho' , \quad \mathbf{v} = \mathbf{v_0} + \mathbf{v'} ,$$

we can linearize the equations. Furthermore, we ignore the mean flow $\mathbf{v_0}$. It is, however, not immediately clear if this is an acceptable simplification, because the mean flow, although much slower than the sound speed c, is not necessarily much slower than the acoustic particle velocity $\mathbf{v'}$. In fact, both are of the order of a few meters per second.

The reason why we indeed can ignore $\mathbf{v_0}$ is because the (small) mean flow only affects the sound field if vorticity is injected from the wall by separation at an edge. In that case the coupling with the vorticity may result in either a source or a sink of acoustic energy. This source or sink is, however, only of importance if the Strouhal number $\omega b/v_0$, based on circular frequency ω and radius of curvature b of the edge, is of order 1 ([7]). In the present problem we expect no appreciable separation at the duct inlet, and as far as separation occurs at the obstacle, a typical Strouhal number very much larger than 1.

As a result, if we ignore v_0 and assume the external force to be stationary with $\nabla p_0 = \mathbf{f}$, we obtain

$$\rho_0 c_0^2 \nabla \cdot \left(\frac{1}{\rho_0} \nabla p' \right) - \frac{\partial^2}{\partial t^2} p' = 0 , \tag{3.16}$$

$$\rho_0 \frac{\partial}{\partial t} \mathbf{v}' + \nabla p' = 0 , \tag{3.17}$$

with $c_0^2 = \gamma R T_0 = \gamma p_0 / \rho_0$. Note that, other than in the constant- c_0 case, the equations would have been different if written in ρ', since $p'_t = c_0^2 (\rho'_t + \mathbf{v}' \cdot \nabla \rho_0)$ (eq. (3.13)).

Since gravity will be negligible here, we have $p_0 = $ constant, and (3.16) can be simplified further. Moreover, since we will consider the behaviour of a single frequency wave it is convenient to introduce

$$p'(\mathbf{x}, t) = \text{Re} \left(p(\mathbf{x}) e^{i\omega t} \right)$$

$$\mathbf{v}'(\mathbf{x}, t) = \text{Re} \left(\mathbf{v}(\mathbf{x}) e^{i\omega t} \right)$$

with circular frequency ω, and $p(\mathbf{x})$ and $\mathbf{v}(\mathbf{x})$ complex functions satisfying

$$\nabla \cdot \left(\frac{1}{k^2} \nabla p \right) + p = 0 \tag{3.18}$$

$$i\omega \rho_0 \mathbf{v} + \nabla p = 0 \tag{3.19}$$

and $k(\mathbf{x}) = \omega / c_0(\mathbf{x})$. The equations (3.18) and (3.19) will be the basis of the analysis to follow.

3.2. Geometry

We will consider a cylindrical hard-walled pipe, in cylindrical coordinates (x, r, θ) given by $r = a$ $(a \sim 0.75$ m$)$ (Figure 1). At $r = a$

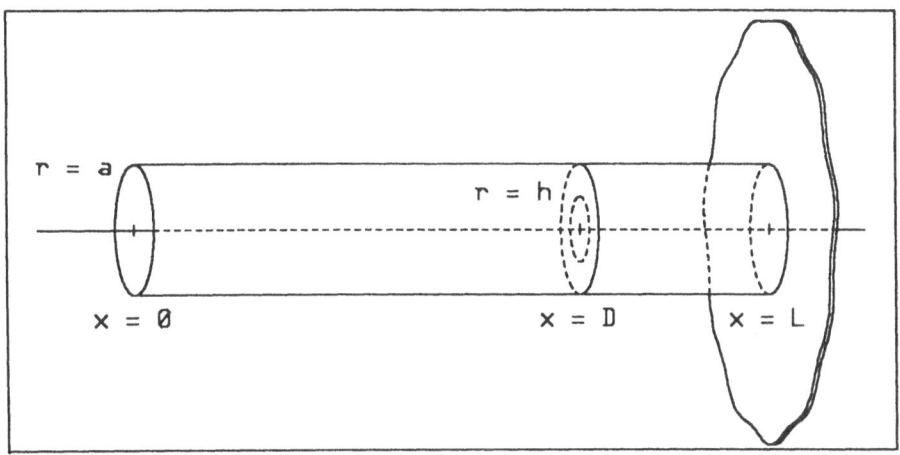

Figure 1. Model geometry

the normal velocity vanishes, $(\mathbf{v} \cdot \mathbf{n}) \sim \partial p / \partial r = 0$. At $x = 0$ a source generates a sound field of circular frequency ω $(\omega \sim 2000/s)$. The frequency is assumed to be so low that only the plane wave is propagating, and no details of the source are to be known. (This requires typically a wavelength larger than a duct diameter.) The part of the pipe $x < 0$ is irrelevant. At $x = L$ $(L \sim 10$ m$)$ the pipe is connected via a flanged opening to the half space $x > L$. The sound speed variation of the mean flow is assumed to be radially symmetric: $c_0 = c_0(x, r)$, $k = k(x, r)$, and constant in $x > L$. Also for the scattering obstacle there is no need to unnecessarily complicate the problem, and we assume an annular hard walled iris at $x = D$ between $r = h$ and $r = a$.

One observation is important and will be utilized in the analysis: although the variation in r of the temperature is rather steep (from 500 K at the wall to $1000 - 2000$ K at the center) and cannot be ignored or otherwise simplified, the variation in x-direction is relatively smooth. It appears that the ratio between a typical wavelength λ and the typical length (L) associated to substantial variations in k or c_0 is small. This parameter, which we will denote by

$\varepsilon = \lambda/L$ $(\varepsilon \sim 0.1 - 0.2)$, suggests an approach of solution where locally the sound speed is constant in x and the field can be described by a modal expansion. On the larger scale this is to be corrected by slowly varying modal amplitudes and wave numbers.

Where it facilitates the analysis, we will make this slow variation explicit by writing $k = k(\varepsilon x, r)$ without using any special notation.

4. SOLUTION

4.1. Modal expansion

When c_0 is independent of x, equation (3.18) is separable in the cylindrical coordinates (x, r, θ). That means that there exist solutions $p(x, r, \theta) = F(x) \, G(r) \, H(\theta)$, satisfying a uniform boundary condition at the coordinate surface $r = a$. These solutions are called modes [4,5]. Mathematically, these modes are interesting because they form a complete basis by which any other solution can be represented by a so-called modal expansion. Physically, these modes are interesting because the usually complicated behaviour of a general field is easier understood via the simpler properties of its modal elements.

If c_0 is constant the modes of a hardwalled infinite duct are the well-known products of Bessel function and exponential functions

$$d_{m\mu}^{\pm}(x, r, \theta) = J_m(\alpha_{m\mu} r) \, e^{\mp i k_{m\mu} x - i m \theta} \tag{4.1}$$

where J_m is the m-th order ordinary Bessel function of the 1st kind [3], $\alpha_{m\mu} a = j'_{m\mu}$ is the μ-th nonnegative nontrivial zero of J'_m, $k_{m\mu} = \sqrt{k^2 - \alpha_{m\mu}^2}$ is the axial wave number with Re $(k_{m\mu}) \geq 0$, Im $(k_{m\mu}) \leq 0$, $\mu = 1, 2, \ldots$, $m = 0, \pm 1, \pm 2, \ldots$.

Since $\alpha_{m\mu}$ grows without limit both for increasing μ and increasing $|m|$, there are only a *finite* number of real $k_{m\mu}$. The rest is purely imaginary. At the right side of a source the $d_{m\mu}^{+}$ modes are generated, of which the ones with real $k_{m\mu}$ are propagating (*cut on*), the other ones with imaginary $k_{m\mu}$ are exponentially decaying (*cut off*). At the left side of a source the same is true for $d_{m\mu}^{-}$.

In the complex plane, the axial wave numbers are typically located as given in Figure 2. A finite number is cut on, between $-k$ on k, and

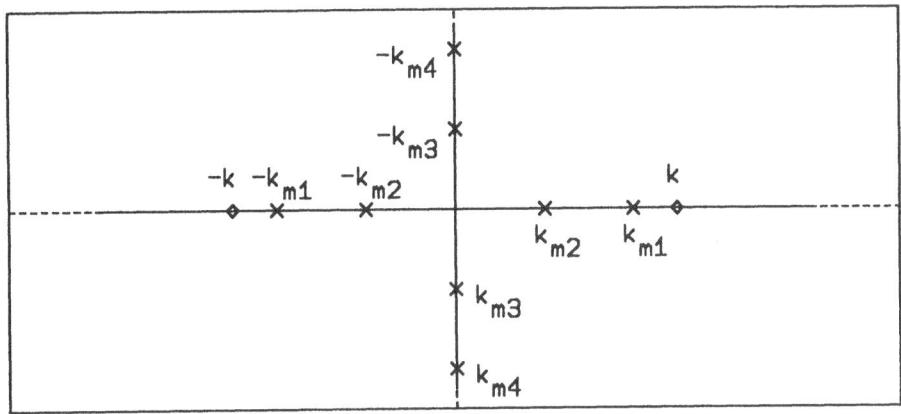

Figure 2. Axial wave numbers.

an infinite number is cut off, along the imaginary axis. It is important to note that if we take k (i.e., ω) small enough, there is only *one* (two if we count left- and right-running separately) mode propagating. This is just what we want in the present problem where we don't want to know details of the source: the other possibly generated modes are exponentially decaying and quickly negligible.

A general solution may be built from the modes $d_{m\mu}^{\pm}$ as the sum (with amplitudes to be determined)

$$p(x, r, \theta) = \sum_{m=-\infty}^{\infty} \sum_{\mu=1}^{\infty} \left(A_{m\mu} d_{m\mu}^{+} + B_{m\mu} d_{m\mu}^{-} \right) . \tag{4.2}$$

Obviously, at the right side of any source or scattering object we have only right-running waves ($B_{m\mu} = 0$), and at the left side only left-running waves ($A_{m\mu} = 0$). Furthermore, if the field is radially symmetric the double series simplifies to a single series for $m = 0$ only.

In the case of a soundspeed depending on r (but *not* on x or θ) the above theory is qualitatively the same. Only the Bessel function becomes now a slightly more general function $\psi_{m\mu}(r)$, defined by the Sturm-Liouville type eigenvalue problem [3]

$$\mathcal{L}_m(\psi; \gamma) = \frac{k^2}{r} \frac{d}{dr} \left(\frac{r}{k^2} \frac{d}{dr} \psi \right) + \left(k^2 - \gamma^2 - \frac{m^2}{r^2} \right) \psi = 0 \tag{4.3}$$

$$|\psi(0)| < \infty , \quad \partial \psi(a)/\partial r = 0 \tag{4.4}$$

where $k_{m\mu} = \gamma$ and $\psi_{m\mu} = \psi$. According to the standard theory, the eigenvalues γ^2 form a real sequence, monotonically tending to $\to -\infty$. The eigenfunctions $\psi_{m\mu}$ form a basis orthogonal to the inner product

$$(\psi_{m\mu}, \psi_{m\nu}) = \int_0^a \psi_{m\mu} \psi_{m\nu} \, r k^{-2} \, dr . \tag{4.5}$$

$\psi_{m\mu}$ has exactly $\mu - 1$ zeros on $(0, a)$. By integrating (4.3) we find for the 1st eigenvalue

$$\gamma_{m1}^2 = \frac{\int_0^a (1 - m^2/k^2 r^2) \, r \, \psi_{m1} \, dr}{\int_0^a \psi_{m1} \, r k^{-2} \, dr} .$$

Since ψ_{m1} has no zeros, $\gamma_{01}^2 > 0$, whereas γ_{m1}^2 for $m \neq 0$ is only positive if k is large enough compared to m.

4.2. Slowly varying amplitude and wave number

In the present problem we cannot have a solution built up from modes because the sound speed c_0 varies in x. However, we can borrow the idea of modes. Since c_0 varies relatively slowly in x the solution is locally representable by an approximate modal expansion. For suitably defined modes such a modal representation then remains the same for all x; only modal amplitude, shape and wave number vary slowly with x [8,9].

Since we are primarily interested in the symmetric problem, we consider from here on only $m = 0$, and do not mention the m-dependence.

For the sake of demonstration it is necessary to make the "slow variation" explicit, and we introduce

$$X = \varepsilon x \tag{4.6}$$

where $k = k(X, r)$, $c_0 = c_0(X, r)$, $\rho_0 = \rho_0(X, r)$.

The slowly varying mode is then assumed to have the form

$$p(x, r) = \psi(X, r; \varepsilon) \, \exp\left(-i \int_0^x \gamma(\varepsilon\xi; \varepsilon) \, d\xi\right) . \tag{4.7}$$

Since the x-derivatives are given by

$$p_x = (-i\gamma\psi + \varepsilon\psi_X) \, \exp\left(-i \int \gamma \, d\xi\right)$$

$$p_{xx} = (-\gamma^2\psi - i\varepsilon\gamma_X\psi - 2i\varepsilon\gamma\psi_X + \varepsilon^2\psi_{XX}) \, \exp\left(-i \int \gamma \, d\xi\right)$$

equation (3.18) becomes

$$\mathcal{L}(\psi; \gamma) = i\varepsilon k^2 \psi^{-1}(\gamma\psi^2 k^{-2})_X - \varepsilon^2 k^2 (\psi_X k^{-2})_X \tag{4.8}$$

with $\psi(r = 0) < \infty$, $\partial\psi(r = a)/\partial r = 0$, where we collected certain groups for later convenience. Since ε is small we expand

$$\psi(X, r; \varepsilon) = \psi_0(X, r) + \varepsilon\psi_1(X, r) + \mathcal{O}(\varepsilon^2)$$

$$\gamma(X; \varepsilon) = \gamma_0(X) + \mathcal{O}(\varepsilon^2)$$

and ψ_0 will be found to be a satisfactory approximation for ψ. We find to leading order

$$\mathcal{L}(\psi_0; \gamma_0) = 0 \tag{4.9}$$

which indeed corresponds to the eigenvalue problem (4.3,4), with now x as a parameter, and solutions ψ_μ, γ_μ (Im $\gamma_\mu \leq 0$). The problem that remains is the way ψ_μ varies with x (the amplitude of ψ_μ is still undetermined, but should vary with x if the energy content of a single mode is to remain the same). For this we consider the equation for perturbation ψ_1:

$$\mathcal{L}(\psi_1; \gamma_0) = ik^2 \psi_0^{-1} (\gamma_0 \psi_0^2 k^{-2})_x .$$ (4.10)

We don't have to solve for ψ_1. It is sufficient to make use of the self-adjointness properties of $\mathcal{L}$, and integrate

$$\int_0^a \psi_1 \mathcal{L}(\psi_0; \gamma_0) \, r k^{-2} dr = \int_0^a \psi_0 \mathcal{L}(\psi_1; \gamma_0) \, r k^{-2} dr$$

$$= i \int_0^a (\gamma_0 \psi_0^2 k^{-2})_x \, r \, dr = 0$$

$$\gamma_0 \int_0^a \psi_0^2 r k^{-2} \, dr = \text{constant}$$ (4.11)

to find (4.11), the equation defining ψ_0 as a function of x. For convenience we take here the constant equal to 1 or $-i$ and obtain

$$(\psi_\mu, \psi_\nu) = |\gamma_\mu|^{-1} \delta_{\mu\nu} .$$ (4.12)

A very important conclusion to be drawn from equation (4.11) or (4.12) is that the theory is invalid and the mode becomes singular at any position x where $\gamma_\mu = 0$. (This may occur for any eigenvalue other than γ_{01}.) This is to be interpreted as follows. If the soundspeed c_0 varies along the duct in such a way that γ_μ vanishes at, say, $x = x_0$ (which depends on both c_0 and frequency ω), then the mode is locally in resonance and changes from propagating (cut on) into decaying (cut off). Since the energy is conserved the mode cannot just disappear but reflects into its backrunning counterpart. These are the spurious unwanted reflections mentioned in chapter 2. They may, as a result, interfere with the reflections from the obstacle and confuse the signal. Therefore, it is important to select a frequency without a resonance for any mode (also $m \neq 0$) along the whole interval $(0, L)$. In that case we do not have to include this turning-point behaviour in the theory [3].

Since, by assumption, there is no interaction between the modes along the smoothly varying interval $0 \leq x \leq D$, we have the general (approximate) solution

$$p(x, r) = \sum_{\mu=1}^{\infty} A_\mu \psi_\mu(X, r) \, \exp\left(-i \int_0^x \gamma_\mu(\varepsilon\xi) \, d\xi\right)$$

$$+ B_\mu \psi_\mu(X, r) \, \exp\left(i \int_0^x \gamma_\mu(\varepsilon\xi) \, d\xi\right) .$$

It is notationally convenient to introduce

$$A_\mu(x) = A_\mu \, \exp\left(-i \int_0^x \gamma_\mu(\varepsilon\xi) \, d\xi\right)$$

$$B_\mu(x) = B_\mu \, \exp\left(i \int_0^x \gamma_\mu(\varepsilon\xi) \, d\xi\right)$$

(4.13)

and similarly along $D \leq x \leq L$ for right running modal amplitudes C_μ and left running D_μ. Then we have

$$p(x, r) = \sum_{\mu=1}^{\infty} \left(A_\mu(x) + B_\mu(x)\right) \psi_\mu(X, r) \quad \text{for} \ \ 0 \leq x < D , \qquad (4.14)$$

$$p(x, r) = \sum_{\mu=1}^{\infty} \left(C_\mu(x) + D_\mu(x)\right) \psi_\mu(X, r) \quad \text{for} \ \ D < x < L . \qquad (4.15)$$

$A_\mu(0)$ are given (the source); $B_\mu(0)$ are to be found; at $x = D$ the incident $A_\mu(D)$ and $D_\mu(D)$ are scattered by the annular obstacle into $B_\mu(D)$ and $C_\mu(D)$; at $x = L$ the incident $C_\mu(L)$ are reflected into $D_\mu(L)$. So to find $B_\mu(0)$ we have to combine reflection and transmission properties of the obstacle at $x = D$ and the open end at $x = L$.

4.3. Scattering by annular obstacle

For a given set of modes incident from $x < D$ (A-modes) and modes incident from $x > D$ (D-modes) we want to know the reflected and transmitted modes generated in $x < D$ (B-modes) and in $x > D$ (C-modes). If we identify with the modal amplitudes $A_\mu(D)$ the vector $\mathbf{A}(D)$, and similarly $\mathbf{B}(D)$, $\mathbf{C}(D)$ and $\mathbf{D}(D)$, this relation is most conveniently expressed by a reflection matrix R and a transmission matrix T.

$$\mathbf{B} = R\mathbf{A} + T\mathbf{D}$$

$$\mathbf{C} = T\mathbf{A} + R\mathbf{D} \ . \tag{4.16}$$

(Due to symmetry, reflection and transmission from the left is the same as from the right.)

A natural method to determine R and T is the technique of mode matching. By projecting the conditions of continuity along an interface to a suitable modal basis (i.e., taking inner products) the problem may be reduced to one of linear algebra. This method is well-known, also for the present iris problem. However, a rather subtle detail in the numerical realisation is only relatively recently well understood [10]. To make this point clear, we will work right from the start with *truncated* series, and assume that our solution will be represented by N modes.

Since the problem is linear it is sufficient to determine the scattered field of a single mode. So we have for $n = 1, \ldots, N$ at

$$x = D- \ : \ p \ = \sum_{\mu=1}^{N} (\delta_{\mu n} + R_{\mu n}) \psi_\mu \ ,$$

$$ip_x = \sum_{\mu=1}^{N} (\delta_{\mu n} - R_{\mu n}) \gamma_\mu \psi_\mu \ , \tag{4.17}$$

$$x = D+ \ : \ p \ = \sum_{\mu=1}^{N} T_{\mu n} \psi_\mu \ ,$$

$$ip_x = \sum_{\mu=1}^{N} T_{\mu n} \gamma_\mu \psi_\mu \ , \tag{4.18}$$

for matrices $R_{N \times N}$ and $T_{N \times N}$. For the moment, the circular area $0 \leq r < h$ at $x = D$ is considered as a duct of length $= 0$, with a set of modes $\hat{\psi}_\mu$ defined by exactly the same equation (4.3) as for $\psi_\mu(r, x = D)$ but now with boundary conditions (4.4) applied at $r = h$. The corresponding inner product will be denoted by

$$[\hat{\psi}_\mu, \hat{\psi}_\nu] = \int\limits_0^h \hat{\psi}_\mu \hat{\psi}_\nu r \, k(r, D)^{-2} \, dr \sim \delta_{\mu\nu} \, . \tag{4.19}$$

We have now in the iris a representation by Q modes,

$$x = D \; : \; ip_x = \sum_{\mu=1}^{Q} U_{\mu n} \hat{\psi}_\mu \text{ on } 0 \leq r \leq h \, ,$$

$$ip_x = 0 \qquad \text{on } h \leq r \leq a \tag{4.20}$$

where we introduced an auxiliary matrix $U_{Q \times N}$.

It is reasonable to take in the iris the number of modes, Q, smaller than N, the number in the full duct. Since the number of zeros of the μ-th mode is $\mu - 1$, the typical radial wavelength of ψ_μ is $(\mu - 1)/2a$ and of $\hat{\psi}_\mu$ $(\mu - 1)/2h$ (see Figure 3).

So a balance between representation accuracies in the iris and the full duct requires about the same smallest wave lengths, or

$$Q = [Nh/a] \, . \tag{4.21}$$

Indeed, this choice yields the fastest convergence for $N \to \infty$ to the physical solution we are interested in. As an example, this behaviour is illustrated in Figure 4.

In addition, it should be noted, that if we take N/Q very much different from a/h, we *may* converge to another solution. This is not an artefact of the method. The problem stated has indeed a non-unique solution. The additional condition necessary to select the correct solution is the so-called *edge condition* [10]: the integrated energy of the field in a neighbourhood of the edge $r = h$ must be

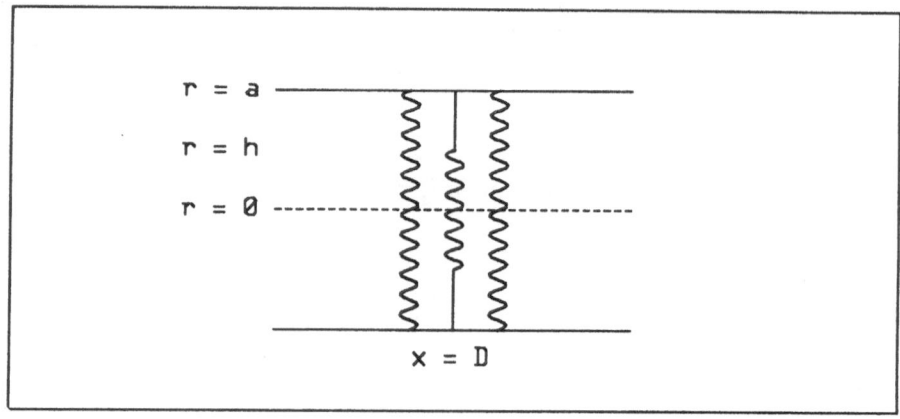

Figure 3. Mode matching through iris.

finite (no source hidden in the edge). Therefore, we may conclude that for the edge condition we should take N/Q not too far from a/h; for reasons of efficiency it is best to take N/Q near a/h as close as possible (eq. (4.21)).

Since p_x is continuous along the *full* interval $0 \le r \le a$ ($p_x = 0$ on both sides of the obstacle along $h \le r \le a$), we have immediately from (4.17) and (4.18) that

$$\delta_{\mu\nu} - R_{\mu\nu} = T_{\mu\nu} \tag{4.22}$$

or in matrix notation $I_{N\times N} - R_{N\times N} = T_{N\times N}$. Using this in the condition of p continuous along $0 \le r \le h$ we find

$$\sum_{\nu=1}^{N} \delta_{\nu n}\psi_\nu = \sum_{\nu=1}^{N} T_{\nu n}\psi_\nu \quad \text{for } 0 \le r \le h \,.$$

Multiply left- and right-hand side with $\hat{\psi}_\mu r/k^2$ and integrate, to obtain

$$[\hat{\psi}_\mu, \psi_n] = \sum_{\nu=1}^{N} [\hat{\psi}_\mu, \psi_\nu] T_{\nu n} \,.$$

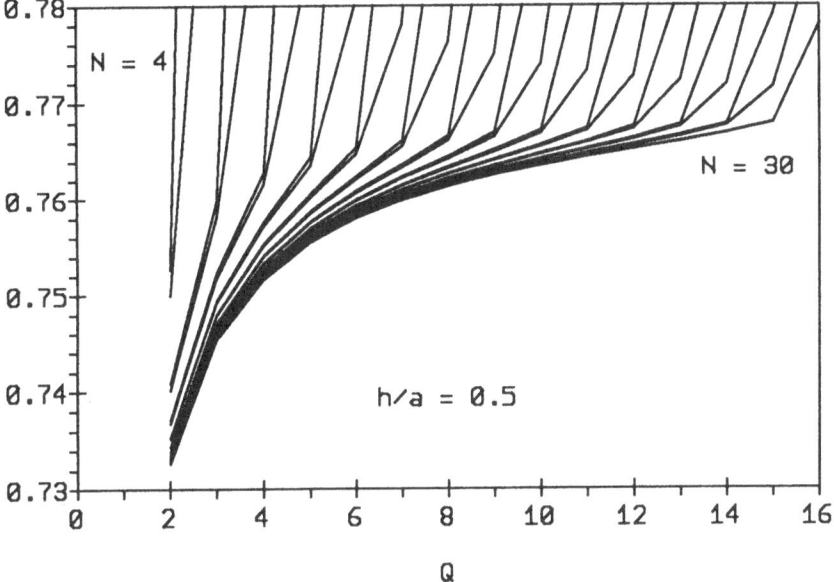

Figure 4. Re (T_{11}) depending on N for various N/Q rates.

If we do this for $\mu = 1, \ldots, Q$ then we have for the auxiliary matrix $S_{Q \times N}$ with elements $S_{\mu\nu} = [\hat{\psi}_\mu, \psi_\nu]$

$$S_{Q \times N} = S_{Q \times N} T_{N \times N} . \tag{4.23}$$

(Note that this implies that $T_{N \times N}$ has at least Q eigenvalues equal to 1.) Finally, from equality of p_x in (4.18) and (4.20) we have

$$\sum_{\nu=1}^{N} T_{\nu n} \gamma_\nu \psi_\nu = \sum_{\nu=1}^{Q} U_{\nu n} \hat{\psi}_\nu \text{ on } 0 \le r \le h$$

$$= 0 \qquad \text{on } h < r \le a .$$

Multiply left- and right-hand side with $\psi_\mu \, r/k^2$ and integrate, to obtain

$$T_{\mu n} \gamma_\mu (\psi_\mu, \psi_\mu) = \sum_{\nu=1}^{Q} [\psi_\mu, \hat{\psi}_\nu] \, U_{\nu n} .$$

For the auxiliary matrix $M_{N \times Q}$ with elements

$$M_{\mu\nu} = [\psi_\mu, \hat{\psi}_\nu] / \gamma_\mu (\psi_\mu, \psi_\mu)$$

we have then

$$T_{N \times N} = M_{N \times Q} \, U_{Q \times N} \, . \tag{4.24}$$

Combining (4.23) with (4.24) and eliminating U gives the final result

$$T_{N \times N} = M_{N \times Q} (S_{Q \times N} \, M_{N \times Q})^{-1} \, S_{Q \times N} \, . \tag{4.25}$$

4.4. Reflection at open end

The reflection of a sound field in a semi-infinite pipe at a flanged open end is (at least for the uniform sound speed as we assumed here in $x \geq L$) classic and goes back to Rayleigh [4,5].

As in the previous chapter the reflection of incident modes $\mathbf{C}(L)$ into backwards running modes $\mathbf{D}(L)$ is most conveniently expressed by an end reflection matrix E as

$$\mathbf{D} = E \, \mathbf{C} \, . \tag{4.26}$$

This matrix E is found as follows.

Using the free field Green's function and its image in the surface $x = L$ we can express the field p in $x > L$ in terms of a normal velocity distribution at the pipe end cross section. For the field just in the pipe opening this *Rayleigh integral* becomes

$$p(L, r) = -\frac{1}{2\pi} \int_0^{2\pi} \int_0^a p_x(L, r') \frac{e^{-ik_0\sigma}}{\sigma} \, r' dr' \, d\theta' \tag{4.27}$$

where $k_0 = k(x \geq L, r)$ (constant) and $\sigma^2 = r^2 + r'^2 - 2rr' \cos(\theta - \theta')$. Since at $x = L$ ψ_μ is now just a multiple of a Besselfunction (equation (4.1) with normalization (4.12)) we may write

$$p(L, r) = \sum_{\mu=1}^{\infty} (C_\mu(L) + D_\mu(L)) \, \psi_\mu(L, r) = \sum_{\mu=1}^{\infty} q_\mu \, J_0(\alpha_{0\mu} r) , \qquad (4.28)$$

$$q_\mu = (C_\mu(L) + D_\mu(L)) \, n_\mu , \quad n_\mu = \sqrt{2} \, k_0/a \, J_0(j'_{0\mu}) \, |k_0^2 - \alpha_{0\mu}^2|^{1/2} ,$$

and similarly

$$p_x(L, r) = -\frac{i}{a} \sum_{\mu=1}^{\infty} v_\mu \, J_0(\alpha_{0\mu} r) , \qquad (4.29)$$

$$v_\mu = a(C_\mu(L) - D_\mu(L)) \, \gamma_\mu n_\mu .$$

Substitute (4.28) and (4.29) into (4.27), and introduce Sonine's integral

$$\frac{e^{-ik_0 \sigma}}{\sigma} = \frac{-i}{a} \int_0^{\infty} \frac{z \, J_0(z\sigma/a)}{w(z)} \, dz , \qquad (4.30)$$

$$w(z) = \sqrt{(k_0 a)^2 - z^2} \quad \text{with} \quad \text{Im}\,(w) \le 0 ,$$

and Gegenbauer's addition theorem

$$J_0(z\sigma/a) = \sum_{m=-\infty}^{\infty} e^{im(\theta - \theta')} \, J_m(zr/a) \, J_m(zr'/a) , \qquad (4.31)$$

so that the r'- and θ'-integrals can be evaluated. We arrive at the equation

$$\sum_{\mu=1}^{\infty} q_\mu \, J_0(\alpha_{0\mu} r) = - \sum_{\mu=1}^{\infty} v_\mu \, J_0(j'_{0\mu}) \int_0^{\infty} \frac{z^2 \, J_0(zr/a) \, J_0'(z)}{w(z) \, (z^2 - j_{0\mu}'^2)} \, dz . \qquad (4.32)$$

Multiply left- and right-hand side with $J_0(\alpha_{0\nu} r)\,r$ and integrate over $0 \le r \le a$ (note orthogonality). Substitute for q_μ and v_μ the original C_μ and D_μ. Then we obtain finally

$$C_\mu(L) + D_\mu(L) = \sum_{\nu=1}^\infty Z_{\mu\nu}(C_\nu(L) - D_\nu(L)) \tag{4.33}$$

with

$$Z_{\mu\nu} = 2a\gamma_\nu \frac{n_\nu J_0(j'_{0\nu})}{n_\mu J_0(j'_{0\mu})} \int_0^\infty \frac{z^3 J'_0(z)^2}{w(z)(z^2 - j'^2_{0\mu})(z^2 - j'^2_{0\nu})} \, dz \; .$$

In matrix notation this is $C + D = Z(C - D)$, or $D = (Z + I)^{-1}(Z - I)C$. Hence, the reflection matrix is

$$E = (Z + I)^{-1}(Z - I) \; . \tag{4.34}$$

(of course, finally to be truncated to $N \times N$).

Finally, we include an additional reflection matrix to matrix E to allow for a possible discontinuous temperature across $x = L$. The proper transition conditions used are continuity of pressure ($[p] = 0$) and momentum ($[k^{-2}p_x] = 0$). For the present purposes this is a minor detail, and will not be further worked out here.

4.5. Gathering the pieces

Now that we have prepared the building blocks of our complete solution ((i) propagation in the smooth parts of the duct; (ii) scattering by the annular obstacle; (iii) reflection by the open end) it is relatively straightforward to assemble them to one coupled system of reflecting and transmitting acoustical elements. The coupling becomes especially clear if we retain the matrix notation already introduced, and present the solution as a modal amplitude vector $B(0)$ for given source $A(0)$.

To describe the modal phase shift from $x = 0$ to D and from D to L we introduce the diagonal matrices

$$
\begin{aligned}
&F \text{ with elements } F_{\mu\mu} = \exp\left(-i \int_0^D \gamma_\mu(\xi) \, d\xi\right) \\
&G \text{ with elements } G_{\mu\mu} = \exp\left(-i \int_D^L \gamma_\mu(\xi) \, d\xi\right) \; .
\end{aligned}
\tag{4.35}
$$

Using (4.26) and (4.16) we have now immediately

$$\mathbf{D}(D) = G\,\mathbf{D}(L) = GE\,\mathbf{C}(L) = GEG\,\mathbf{C}(D)$$

$$= GEGT\,\mathbf{A}(D) + GEGR\,\mathbf{D}(D)$$

and so

$$\mathbf{D}(D) = (I - GEGR)^{-1}\,GEGT\,\mathbf{A}(D) = H\,\mathbf{A}(D)\ . \qquad (4.36)$$

Finally, the solution we were looking for is

$$\mathbf{B}(0) = F\,\mathbf{B}(D) = FR\,\mathbf{A}(D) + FT\,\mathbf{D}(D)$$

$$= F(R + TH)\,\mathbf{A}(D)$$

$$\mathbf{B}(0) = F(R + TH)F\,\mathbf{A}(0)\ . \qquad (4.37)$$

Note that although the exponential decay of the cut-off modes would make it possible to consider only one mode in $[0, D]$, this is not the case along $[D, L]$ because this interval may be relatively short.

5. RESULTS

5.1. Introduction

Any sufficiently complex mathematical model of a serious engineering problem has to be implemented, eventually, as a computer program. Of course, trends, simplified cases, the behaviour near singularities in parameter and variable space, and the character of single isolated effects should be understood analytically as much as possible. But the interaction of various equally important components and the rôle of more general configurations and more realistic geometries can only be studied by numerical simulation.

The solution of the present problem is split up in components for which the numerical solution is, although not straightforward, standard in the sense that we can utilize the world-wide available well-tested robust public-domain software [11], or routines of commercial libraries of numerical software like NAG [12] or IMSL [13].

The first important problem to be solved numerically is the Sturm-Liouville eigenvalue problem (4.3) and (4.4). In particular the eigenvalues are essential, because we want to design the configuration such that for $m = 0$ only the first eigenvalue γ_{01} is real, and that for $|m| \geq 1$ all eigenvalues are imaginary.

If the temperature (and therefore $k(x, r)$) is independent of r, the solution is just the Bessel function of the first kind J_m, so this can serve as a check for the numerics. The routine we used is the Fortran translation TSTURM by B.S. Garbow (Argonne National Laboratory) of the Algol procedure Tristurm by Peters and Wilkinson ([14]). Details of the application like accuracy and prediction of the number of eigenvalues are necessary but will not be considered here because these evidently depend on the routine chosen.

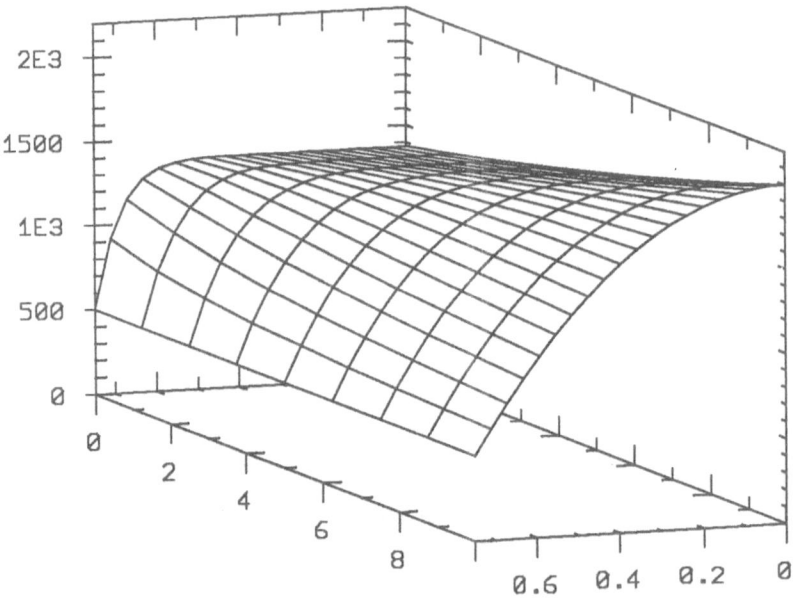

Figure 5. Temperature profile.

For the open end reflection we need to evaluate the integral in each matrix $Z_{\mu\nu}$ element (eq. (4.33)). The convergence rate at infinity of the integrand is rather poor, especially for increasing μ and ν, and we rewrote the integral by using complex contour deformation into a set of numerically more pleasant integrals. One Bessel function is rewritten as a sum of Hankel functions: $J_0' = \frac{1}{2}H_0^{(1)'} + \frac{1}{2}H_0^{(2)'}$, the first of which converges in the upper complex half plane, and the other in the lower half plane. The integral may subsequently be split up in two of which the contours can be deformed to the positive and to the negative imaginary axis. After taking care of the branch cut of square root $w(z)$ and possible residues if $\mu = \nu$, we arrive at two integrals which can be evaluated in a standard way (we used Romberg integration).

The matrix and vector manipulations necessary for the scatter-

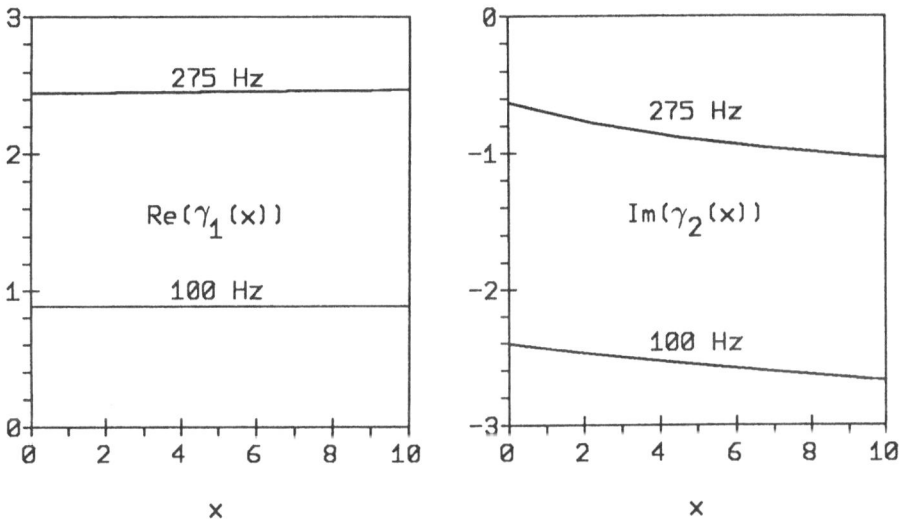

Figure 6. First two modes for $f = 100\ Hz$ and $f = 275\ Hz$.

ing problem (eq. [4.22–25]) and the coupling of the various parts (eq. [4.36,37]) are taken from the BLAS/LINPACK package written by Dongarra et al. ([15]).

5.2. Example
The realistic example we have considered is a duct of length $L =$

10 m, radius $a = 0.75$ m, and a temperature profile as given in Figure 5. The temperature varies along the centerline from 2000 K to 1350 K, and remains along the wall at 500 K, such that the cross sectional average is constant 1250 K. (This is also the temperature in the furnace chamber $x > L$).

We scanned the modal x-variation of two frequencies: $f = 100\ Hz$ and $f = 275\ Hz$. One much lower and the other just low enough for the first mode to be the only one cut-on (Figure 6). Note that the first few modes appear to vary only very slightly in x. This is caused by the temperature varying such that the average is constant. As a result only a few positions in x are necessary.

The most important question for the engineer involved with the

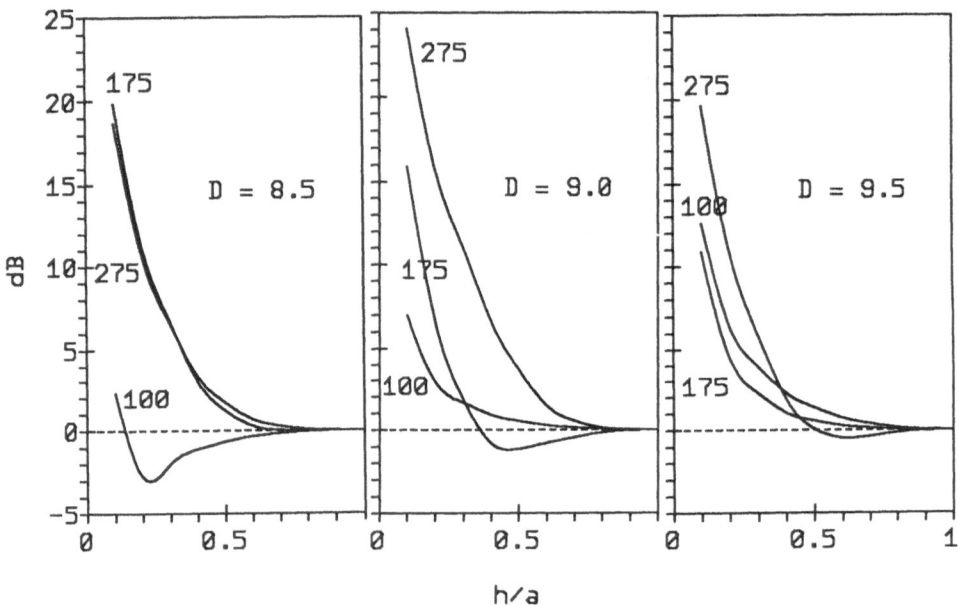

Figure 7. Relative transmitted power Δ for $D = 8.5$, 9, 9.5 m at $f = 100,\ 175,\ 275\ Hz$.

contamination problem is of course how much of a constriction of the pipe can be observed from the acoustical reflection. To this end we varied the iris radius h for three frequencies (Figure 7), and the frequency

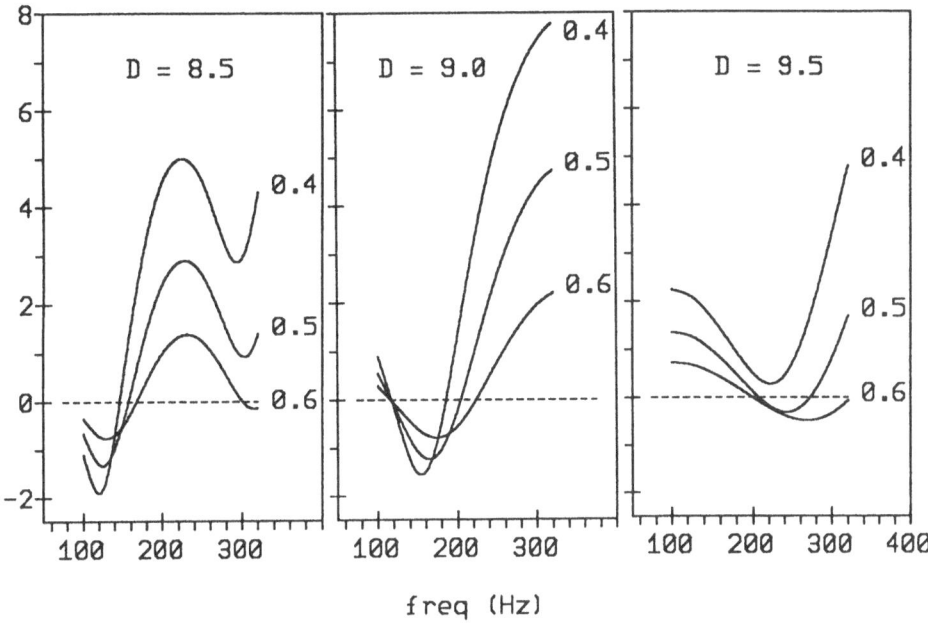

Figure 8. Relative transmitted power Δ for $D = 8.5$, 9, 9.5 m at $h/a = 0.4$, 0.5, 0.6.

f for three radii h (Figure 8), at three positions D. We compared the acoustic transmitted power $\mathcal{P} \sim |A(0)|^2 - |B(0)|^2$ [4,5] with ($\mathcal{P}$) and without ($\mathcal{P}_0$) the obstacle. This is commonly expressed in decibels as $\Delta - 10 \log_{10}(\mathcal{P}_0/\mathcal{P})\, dB$. If the reflected signal is strong enough, $\mathcal{P}$ becomes small, and for a difference larger than, say, $\Delta = 3\, dB$ we can safely say that the obstacle can be detected.

We see that in general a high frequency is favourable for the obstacle to be visible. There is, however, always the inopportune possibility of resonance and the excitation of a standing wave, increasing the net transmission so much that with obstacle more energy is transmitted than without ($\Delta < 0$).

This effect may be considered as a typical example of an interaction effect, adverse to intuition, and only found (usually) by numerical experiments.

Of course, what we now should do is to further investigate the influence of position D, temperature profile, and consider in detail the behaviour of the reflection matrices R and E, and the sound pressure profile in r. However, this is a bit outside the scope of the present introductory purposes.

We end our discussion of results here with the remark that it is very important for the applied mathematician to understand the physics of his problem well. Apart from the selection of the model and the numerical methods, this is vital for the interpretation and feed back to the model of the final results.

6. CONCLUSIONS

As an example of a mathematical engineering problem we have considered the acoustical detection of constrictions in a pipe due to contamination by liquid particles of ashes in gasified coal. By applying a semi-analytical solution based on slowly varying modes one can select an acoustic frequency such that there are guaranteed no spurious reflections due to the temperature gradients. Although for relatively low frequencies small obstacles are not visible, a constriction of say 50% is well visible for high enough frequencies. For some frequencies the constriction appeared, in some cases, to reduce the reflection and spoil the detection. It is clear that for any diagnosis a complete frequency range is to be taken into consideration.

7. REFERENCES

1.   D.J. Struik, A concise history of mathematics, Dover, New York, 1948.

2.   H.K. Kuiken, Industrial Mathematics, to appear in: Lecture Notes in Mathematics, Mathematical Modelling of Industrial Processes, Springer, Heidelberg, 1992.

3.   C.M. Bender and S.A. Orszag, Advanced mathematical methods for scientists and engineers, McGraw-Hill, New York, 1978.

4.   P.M. Morse and K.U. Ingard, Theoretical Acoustics, McGraw-Hill, New York, 1968.

5.   A.D. Pierce, Acoustics: an introduction to its physical principles and applications, McGraw-Hill, New York, 1981.

6.   P.A. Thompson, Compressible fluid dynamics, McGraw-Hill, New York, 1972.

7.   M.C. Peters, A. Hirschberg, J.A. v.d. Konijnenberg, F.J. Huijsmans, R.W. de Leeuw, S. Op de Beek, A.P. Wijnands, Experimental study of the aero-acoustic behaviour of an open pipe termination at low Mach numbers and low Helmholtz numbers, DGLR/AIAA 92–02–055, DGLR/AIAA 14th Aeroacoustics Conference, May 11–14, 1992. Aachen, Germany.

8.   A.H. Nayfeh, Introduction to perturbation techniques, John Wiley & Sons, New York, 1981.

9.   S.W. Rienstra, Sound transmission in a slowly varying lined flow duct, Nieuw Archief voor Wiskunde, series 4, part 6, no. 1–2, (1988) 157.

10. R. Mittra and S.W. Lee, Analytical techniques in the theory of guided waves, The MacMillan Company, New York, 1971.

11. BLAS library, LINPACK library, available from Netlib.

12. The Numerical Algorithms Group Ltd., Wilkinson House, Jordan Hill Road, Oxford, England.

13. International Mathematical and Statistical Libraries Inc., Houston, Texas, U.S.A.

14. G. Peters and J.H. Wilkinson, Handbook for Automatic Computation, Vol. 2 – Linear Algebra, 418–439, Springer, Heidelberg, 1971.

15. J.J. Dongarra, C.B. Moler, J.R. Bunch, G.W. Stewart, LINPACK User's Guide, SIAM, Philadelphia, 1979.

Interior point approach to linear programming: theory, algorithms & parametric analysis

C. Roos

Faculty of Technical Mathematics and Informatics

Delft University of Technology

P.O. Box 5031, 2600 GA Delft, The Netherlands

Abstract

The classical theory of linear programming strongly depends on the fact that among the optimal solutions of an LP-problem there is always a vertex solution. In many situations the analysis is complicated by the fact that this vertex solution may not be unique. The recent research in the field of interior point methods for LP has made clear that every (solvable) LP-problem has a unique socalled central solution, namely the analytic center of the optimal facet of the problem. In this paper we reconsider the theory of LP by using central solutions. The analysis is facilitated by the unicity of the central solution of an LP-problem. Starting from scratch, using an elementary result from calculus, we present new proofs of the fundamental results of LP. These include the existence of a strictly complementary solution, and the strong duality theorem for LP. The proofs are simpler and often more natural than the ones currently known. It turns out that the central solution of an LP-problem is the limit point of the socalled central path of the problem. Based on this observation an algorithm will be derived which approximately follows the central path. The output of this algorithm is an approximation of the central solution. It also gives us the optimal partition of the problem. We finally deal with the topic of parametric analysis. So we investigate the dependence of the central solution on the right hand side coefficients and/or the coefficients in the objective vector. It turns out that also from the parametric point of view the interior point approach is more natural than the usual simplex based approach.

1 Introduction

In 1984 Karmarkar [14] discovered an interior point method (IPM) for linear programming (LP) with a polynomial–time complexity bound, and he claimed it to be very efficient in practice. This resulted in tremendous research activity in the field of IPM for LP. As a result many theoretical

181

A. van der Burgh and J. Simonis (eds.), Topics in Engineering Mathematics, 181–216.

© 1992 *Kluwer Academic Publishers.*

improvements to the basic method have been made, and several variants of the method are shown to be computationally efficient, especially for large scale linear programs. The theoretical and practical appeal of IPM's at present is unquestionable.

One deficiency in the current state of affairs is that the IPM's have relied on the existing theory of linear programming. From the theoretical point of view it would be more satisfying if we could derive the theory for linear programming within the context of an IPM approach. Section 2 of this paper is devoted to this subject. We demonstrate that the interior point techniques can be used in a natural and easy manner to establish the duality theorem of linear programming (LP duality theorem). The new proof of the LP duality theorem has the attractive feature that it yields as a side result the existence of a strictly complementary optimal solution pair, which is a well known result of Goldman and Tucker [9]. In Section 3 we proceed by presenting two interior point algorithms with a polynomial–time bound, one for the standard LP–problem and one for its dual problem. Finally, the Section 4 is devoted to the classical, and from the practical point of view very important topic of parametric analysis, now based on the knowledge of a central solution of the problem.

Throughout the paper we use the following notation. If x denotes a vector, then the corresponding capital letter X will denote the diagonal matrix with the components of x on the diagonal. Furthermore, e will always denote a vector of all ones, and E the identity matrix of appropriate size.

2   Central approach to the theory of LP

In this section we start with proving the LP duality theorem and the Goldman–Tucker theorem for (P) under the assumption that both the primal and the dual feasible sets contain interior points. Then we remove the interior point assumption and obtain the same results for the general situation.

In this paper we deal with a linear program in the standard form:

$$(P) \qquad \min\left\{c^T x : Ax = b,\ x \geq 0\right\},$$

where A is an $m \times n$ matrix, $c, x \in \mathbb{R}^n$, and $b \in \mathbb{R}^m$. The linear program (P) has associated with it the dual program

$$(D) \qquad \max\left\{b^T y : A^T y + s = c,\ s \geq 0\right\},$$

with $s \in \mathbb{R}^n$, and $y \in \mathbb{R}^m$. The feasible regions of (P) and (D) are denoted as $\mathcal{P}$ and $\mathcal{D}$ respectively. So

$$\mathcal{P} := \{x \; : \; Ax = b, \, x \geq 0\},$$

$$\mathcal{D} := \{(y, s) \; : \; A^T y + s = c, \, s \geq 0\}.$$

We may assume that A has full rank m, because otherwise some of the equality restrictions in (P) are redundant and hence can be removed. This implies that there exists a one–to–one correspondence between y and s in the pairs $(y, s) \in \mathcal{D}$. In order to facilitate the discussion we shall feel free to refer to any pair $(y, s) \in \mathcal{D}$ either by $y \in \mathcal{D}$ or $s \in \mathcal{D}$. The (relative) interiors of $\mathcal{P}$ and $\mathcal{D}$ are denoted as $\mathcal{P}^0$ and $\mathcal{D}^0$. So we have

$$\mathcal{P}^0 := \{x \; : \; Ax = b, \, x > 0\},$$

$$\mathcal{D}^0 := \{(y, s) \; : \; A^T y + s = c, \, s > 0\}.$$

For future use we recall the well known and almost trivial 'Weak Duality Theorem' for linear programming.

Theorem 2.1 *Let x and s be feasible for (P) and (D) respectively. Then $c^T x - b^T y = x^T s \geq 0$. Consequently, if the 'duality gap' $x^T s$ is zero then x is an optimal solution of (P) and (y, s) is an optimal solution of (D).*

Proof: The proof is straightforward. One has

$$0 \leq x^T s = x^T (c - A^T y) = c^T x - (Ax)^T y = c^T x - b^T y.$$

This implies that $c^T x$ is an upper bound for the optimal objective value of (D), and $b^T y$ is a lower bound for the optimal objective value of (P), and, moreover, if the duality gap is zero then the pair (x, y) is optimal. $\qquad \square$

2.1 LP duality under an interior point assumption

Under the assumption that both the primal and dual feasible regions have interior points, we show that both problems have optimal solutions. For each of the two problems, one of the optimal solutions is the socalled *central* solution, which will be characterized in two ways: as the limit point of the *central path* of the problem, and also as the *analytic center* of the set of optimal solutions.

We shall need the following well–known result from elementary convex analysis. We include its proof for completeness. This result will be used only in the proof of Lemma 2.1. It is possible to circumvent its use at the expense of somewhat tedious calculations. Recall that a subset C of $\mathbb{R}^k$ is called *relatively open*, if C is open in the smallest affine subspace of $\mathbb{R}^k$ containing C.

Theorem 2.2 *Let $f : D \to \mathbb{R}$ be a differentiable convex function, where $D \subseteq \mathbb{R}^k$ is an open set and let C be a relatively open subset of D. Then, $x \in C$ minimizes f over C if and only if*

$$\nabla f(x)^T(z - x) = 0, \quad \forall z \in C. \tag{1}$$

Proof: Since f is convex, we have for any $x, z \in C$

$$f(z) \geq f(x) + \nabla f(x)^T(z - x).$$

The sufficiency of condition (1) follows immediately. To prove the necessity of (1), consider the point $z_t = x + t(z - x)$, $0 \leq t \leq 1$. We have $z_t \in C$ and thus

$$f(z_t) \geq f(x) + t\nabla f(x)^T(z - x).$$

Since $f(z_t) \geq f(x)$, letting $t \to 0$ we have

$$\nabla f(x)^T(z - x) = \lim_{t \to 0} \frac{f(z_t) - f(x)}{t} \geq 0.$$

Now let x minimize f. Then, if $z \in C$ and $w := z - x$, for some positive λ also $z' := x - \lambda w \in C$. Now we have both $\nabla f(x)^T w \geq 0$ and $\nabla f(x)^T(-\lambda w) \geq 0$. Hence (1) follows. □

For any positive number $\mu > 0$, we define $f_\mu : \mathcal{P}^0 \to \mathbb{R}$ by

$$f_\mu(x) := c^T x - \mu \sum_{j=1}^{n} \ln x_j.$$

f_μ is called the logarithmic barrier function for (P) with barrier parameter μ.

Lemma 2.1 *Let $\mu > 0$. The following statements are equivalent:*

(i) There exists a (unique) minimizer of f_μ;

(ii) There exist $x, s \in \mathbb{R}^n$ *and* $y \in \mathbb{R}^m$ *such that*

$$
\begin{aligned}
Ax &= b, & x \geq 0, \\
A^T y + s &= c, & s \geq 0, \\
XS &= \mu E.
\end{aligned}
\tag{2}
$$

Proof: First note that whenever x, y and s solve (2), then both x and s are positive, due to the third equation. So the nonnegativity conditions for x and s in (2) can equally well be replaced by requiring that x and s are positive.

One easily checks that f_μ is strictly convex. Hence there exists at most one minimizer of f_μ over the set $\mathcal{P}^0$. Since the domain of f_μ is relatively open, Theorem 2.2 applies. Thus it follows that f_μ has $x \in \mathcal{P}^0$ as a minimizer if and only if

$$
\begin{aligned}
Ax &= b, & x > 0, \\
(c - \mu X^{-1} e)^T w &= 0, & \forall w \in N(A),
\end{aligned}
$$

where $N(A)$ denotes the null space of A. The second equation is equivalent to saying that $c - \mu X^{-1} e$ lies in the orthogonal complement of $N(A)$. According to a fundamental result in linear algebra the orthogonal complement of $N(A)$ is the row space $R(A^T)$ of A. See e.g., [28, 22]. Thus it follows that $x \in \mathcal{P}^0$ minimizes f_μ if and only if $c - \mu X^{-1} e = A^T y$ for some $y \in \mathbb{R}^m$. Setting $s = \mu X^{-1} e$, the lemma follows. $\square$

Let H be a matrix whose rows form a basis for the subspace $N(A)$. Then, $N(A) = R(H^T)$, $R(A^T) = N(H)$, and $\mathbb{R}^n = R(H^T) \oplus N(H)$ is an orthogonal decomposition of $\mathbb{R}^n$. Now assume that $x^0 \in \mathcal{P}^0$ and $(y^0, s^0) \in \mathcal{D}^0$. Then the first equation in (2) can be written as $Ax = Ax^0$, and, using the matrix H, the second equation as $Hs = Hs^0$. Thus we get the following symmetric form of (2):

$$
\begin{aligned}
Ax &= Ax^0, & x \geq 0, \\
Hs &= Hs^0, & s \geq 0, \\
XS &= \mu E.
\end{aligned}
\tag{3}
$$

(In the context of interior point methods, the symmetric form above is apparently used for the first time in Todd–Ye [30].) Using the symmetry in x and s we can apply Lemma 2.1 to the function $h_\mu : \{s : Hs = Hs^0, s > 0\} \to \mathbb{R}$, defined by

$$
h_\mu(s) := (x^0)^T s - \mu \sum_{j=1}^{n} \ln s_j.
$$

This shows that h_μ has a (unique) minimizer if and only if the system (2) has a solution. Observe that

$$(x^0)^T s = (x^0)^T (c - A^T y) = (x^0)^T c - (x^0)^T A^T y = (x^0)^T c - b^T y. \quad (4)$$

So, defining the logarithmic barrier function for (D) by

$$g_\mu(s) := b^T y + \mu \sum_{j=1}^n \ln s_j,$$

with domain $\mathcal{D}^0$ (the interior of $\mathcal{D}$), we see that g_μ and $-h_\mu$ differ by a constant. Hence g_μ has a maximizer if and only if (2) has a solution. The proof of the following theorem uses this fact as well as a straightforward lemma from calculus which we state without proof.

Lemma 2.2 *Let the function* $f : (0, \infty) \rightarrow \mathbb{R}$ *be defined by* $f(x) := \sigma x - \mu \ln x$, *where* σ *and* μ *are positive. Then* $f(x)$ *is bounded from below. Moreover, for each* $M \in \mathbb{R}$ *there exists an interval* $[a, b]$, *with* $0 < a < b$, *such that* $f(x) \geq M$ *whenever* $x \notin [a, b]$.

Theorem 2.3 *Let* $\mu > 0$. *Then, the following statements are equivalent:*

(i) Both $\mathcal{P}$ *and* $\mathcal{D}$ *contain positive vectors;*

(ii) There exists a (unique) minimizer of f_μ;

(iii) There exists a (unique) maximizer of g_μ;

(iv) The system (2) has a solution.

Proof: The equivalence of *(ii)* and *(iv)* is already contained in Lemma 2.1, and the equivalence of *(iv)* to *(iii)* is proved above. It remains to show that *(i)* is equivalent to *(ii)* – *(iv)*. Earlier we already noted the obvious fact that *(iv)* implies *(i)*. So it suffices for the proof of the theorem if we show that *(i)* implies *(ii)*.

Assuming *(i)*, we have $x^0 \in \mathcal{P}^0$ and $(y^0, s^0) \in \mathcal{D}^0$ such that $x^0 > 0$ and $s^0 > 0$. First observe that minimizing the barrier function f_μ over the set $\mathcal{P}^0$ is equivalent to minimizing the function

$$k_\mu(x) := (s^0)^T x - \mu \sum_{j=1}^n \ln x_j$$

over the same set. This is due to the relation

$$(s^0)^T x = (c - A^T y^0)^T x = c^T x - b^T y_0.$$

So the proof will be complete if we show that k_μ has a minimizer in $\mathcal{P}^0$. To this end we write

$$k_\mu(x) = \sum_{j=1}^n \left(s_j^0 x_j - \mu \ln x_j \right).$$

Due to Lemma 2.2 we have numbers τ_j such that the j-th term in the above sum is bounded from below by τ_j. Moreover, defining

$$M_j := k_\mu(x^0) - (\tau_1 + \ldots + \tau_{j-1} + \tau_{j+1} + \ldots + \tau_n), \ 1 \le j \le n,$$

we have $M_j \ge \tau_j$. Using Lemma 2.2 once more we obtain the existence of intervals $[a_j, b_j]$, $0 < a_j < b_j$, such that the value of the j-th term exceeds M_j whenever $x_j \notin [a_j, b_j]$. So, if $x \in \mathcal{P}^0$ and $x_j \notin [a_j, b_j]$ for some j, then

$$k_\mu(x) \ge \tau_1 + \ldots + \tau_{j-1} + M_j + \tau_{j+1} + \ldots + \tau_n \ge k_\mu(x^0).$$

We conclude from this that the level set $L := \{x \in \mathcal{P}^0 : k_\mu(x) \le k_\mu(x^0)\}$ is a subset of the cartesian product of the intervals $[a_j, b_j]$, $0 \le j \le n$. Hence the level set L is bounded. Note that L is nonempty, because $x^0 \in L$. Since k_μ is continuous, it has a minimizer in L, and consequently also in $\mathcal{P}^0$. This completes the proof.   □

In the remainder of this paragraph, we will make the basic assumption that statement (i) of Theorem 2.3 holds. We shall remove this assumption in the next paragraph.

Assumption 2.1 *Both $\mathcal{P}$ and $\mathcal{D}$ contain a positive vector.*

For each positive μ we will denote the minimizer of f_μ as $x(\mu)$ and the maximizer of g_μ as $(y(\mu), s(\mu))$. The set $\{x(\mu) : \mu > 0\}$ is called the *central path* of (P), and likewise $\{(y(\mu), s(\mu)) : \mu > 0\}$ the *central path* of (D).

Recall from Theorem 2.1 that if $(x, s) \in \mathcal{P} \times \mathcal{D}$, then $x^T s \ge 0$, and if moreover $x^T s = 0$ then x is optimal for (P) and s is optimal for (D). We proceed by showing the existence of a *complementary solution* $(x^*, s^*) \in \mathcal{P} \times \mathcal{D}$, that is, a solution satisfying $(x^*)^T s^* = 0$. We need the following lemma.

Lemma 2.3 *Let $\bar{\mu} > 0$. The set $\{(x(\mu), s(\mu))\}_{0 < \mu \leq \bar{\mu}}$ is bounded.*

Proof: Since $x^0 - x(\mu) \in N(A)$ and $s^0 - s(\mu) \in R(A^T)$, we have $(x^0 - x(\mu))^T(s^0 - s(\mu)) = 0$. For any j, $1 \leq j \leq n$, this implies

$$s_j^0 x_j(\mu) \leq (s^0)^T x(\mu) + (x^0)^T s(\mu) = n\mu + (x^0)^T s^0 \leq n\bar{\mu} + (x^0)^T s^0.$$

This shows $x_j(\mu) \leq (n\bar{\mu} + (x^0)^T s^0)/s_j^0$ so that $\{x(\mu)\}_{0 < \mu \leq \bar{\mu}}$ is bounded. The proof for $\{s(\mu)\}_{0 < \mu \leq \bar{\mu}}$ is similar. $\qquad\square$

We now have the necessary ingredients to prove the LP duality theorem under Assumption 2.1.

Theorem 2.4 *Let x and s be feasible for (P) and (D), respectively. Then the points x and s are optimal solutions if and only if $x^T s = 0$. Moreover, both problems have optimal solutions.*

Proof: By virtue of Theorem 2.1, it suffices to show that there exist optimal solutions x^* for (P) and (y^*, s^*) for (D) such that the duality gap $(x^*)^T s^*$ vanishes. Let $\mu_k \to 0$. Since the set $\{(x(\mu_k), s(\mu_k))\}$ is bounded, it contains a subsequence converging to a point (x^*, s^*). We have $(x^*, s^*) \in \mathcal{P} \times \mathcal{D}$ and, since $x(\mu_k)^T s(\mu_k) = n\mu_k \to 0$, $(x^*)^T s^* = 0$. $\qquad\square$

One early result in the theory of linear programming due to Goldman and Tucker [9] concerns the existence of a *strictly complementary primal–dual optimal solution pair* (see also Schrijver [26], pp. 95–96). This is an optimal solution pair (x^*, s^*) satisfying $x^* + s^* > 0$. As we shall now describe, there is a rich structure associated with these solutions.

We first introduce some notation. Denote by z^* the common optimal value of problems (P) and (D). The sets of optimal solutions for (P) and (D) will be denoted as $\mathcal{P}^*$ and $\mathcal{D}^*$ respectively. We have

$$\begin{aligned}
\mathcal{P}^* &= \{x : Ax = b, c^T x = z^*, x \geq 0\}. \\
\mathcal{D}^* &= \{(y, s) : A^T y + s = c, b^T y = z^*, y \in \mathbb{R}^m, s \geq 0\}.
\end{aligned}$$

If x is any vector in $\mathbb{R}^n$, we define its *support*, denoted as $\sigma(x)$, to be the set of coordinate indices i for which $x_i > 0$. Let $(x, s) \in \mathcal{P} \times \mathcal{D}$. As a consequence of the LP duality theorem (Theorem 2.4), we have $\sigma(x) \cap \sigma(s) = \emptyset$ if and only if (x, s) is an optimal primal–dual pair. Note that (x, s) is a strictly complementary optimal primal–dual pair if and only if $\sigma(x) \cup \sigma(s) = \{1, \ldots, n\}$. We now present a simple proof of the existence of such a pair under Assumption 2.1.

Theorem 2.5 *There exist points $x^* \in \mathcal{P}^*$ and $s^* \in \mathcal{D}^*$ such that (x^*, s^*) is a strictly complementary optimal pair.*

Proof: Let x^* and s^* be as defined in the proof of Theorem 2.4. Thus we have a sequence $\{(x(\mu_k), s(\mu_k))\}$, with $\mu \to 0$, which converges to (x^*, s^*). We claim that (x^*, s^*) is a strictly complementary pair. The proof goes as follows.

Since $x(\mu_k) - x^* \in N(A)$ and $s(\mu_k) - s^* \in R(A^T)$, we have

$$(x(\mu_k) - x^*)^T (s(\mu_k) - s^*) = 0.$$

Rearranging the terms of this equality, and noting that $x(\mu_k)^T s(\mu_k) = n\mu_k$, $(x^*)^T s^* = 0$, we arrive at

$$\sum_{j \in \sigma(x^*)} x_j^* s_j(\mu_k) + \sum_{j \in \sigma(s^*)} x_j(\mu_k) s_j^* = n\mu_k.$$

Dividing both sides by μ_k and recalling that $x_j(\mu_k) s_j(\mu_k) = \mu_k$, we obtain

$$\sum_{j \in \sigma(x^*)} \frac{x_j^*}{x_j(\mu_k)} + \sum_{j \in \sigma(s^*)} \frac{s_j^*}{s_j(\mu_k)} = n.$$

Letting $k \to \infty$, we see that the first sum above becomes equal to the number of nonzero coordinates in x^*. Similarly, the second sum becomes equal to the number of nonzero coordinates in s^*. We conclude that the pair (x^*, s^*) is strictly complementary. $\square$

With x^* and s^* in Theorem 2.5, we partition the index set $\{1, \ldots, n\}$ according to

$$\{1, \ldots, n\} = B \cup N, \tag{5}$$

with

$$B := \{i : x_i^* > 0\},$$

$$N := \{i : s_i^* > 0\}.$$

Theorem 2.6 *The partition given in (5) is unique. In other words, every strictly complementary optimal pair (x, s) has the same support: $\sigma(x) = B$ and $\sigma(s) = N$.* $\square$

Proof: As a consequence of Theorem 2.4, $x \in \mathcal{P}$ is optimal if and only if $x^T s^* = 0$. Similarly, $s \in \mathcal{D}$ is optimal if and only if $(x^*)^T s = 0$. Hence it follows that

$$x \in \mathcal{P}^* \quad \Leftrightarrow \quad \sigma(x) \subseteq B,$$
$$s \in \mathcal{D}^* \quad \Leftrightarrow \quad \sigma(s) \subseteq N. \qquad \square$$

The partition (5) of the index set $\{1, \ldots, n\}$ will be called the *optimal partition* of the problem (P) and of the problem (D).

We will use the notation x_B and x_N to refer to the restriction of a vector $x \in \mathbb{R}^n$ to the coordinate sets B and N respectively. The submatrices A_B and A_N of a matrix A are defined in the same way. As a consequence of Theorem 2.6, we have

$$\mathcal{P}^* = \{x : A_B x_B = b, \ x \geq 0, \ x_N = 0\},$$
$$\mathcal{D}^* = \{(y, s) : A^T y + s = c, \ y \in \mathbb{R}^m, \ s \geq 0, \ s_B = 0\}.$$

The following result immediately follows from the descriptions of $\mathcal{P}^*$ and $\mathcal{D}^*$ above.

Corollary 2.1 *The set of strictly complementary optimal solutions coincide with the (relative) interior of the optimal faces $\mathcal{P}^*$ and $\mathcal{D}^*$.* $\qquad \square$

An interesting observation now leads us to the notion of *the central solution* of an LP–problem. Every optimal solution x of (P) satisfies $x_N = 0$, and there exists at least one optimal solution such that $x_B > 0$. Due to the boundedness of the optimal face the following maximization problem

$$\max \left\{ \sum_{i \in B} \ln x_i \ : \ x \in \mathcal{P}^* \right\} \tag{6}$$

has an optimal solution. Since the objective function is strictly convex, as can easily be verified, the optimal solution is unique. This is the so–called *analytic center* of $\mathcal{P}^*$ (see Sonnevend [27]). We will refer to this optimal solution of (P) as the *central* solution of (P). The *central* solution of (D) is defined in a similar way, namely as the unique solution of the optimization problem

$$\max \left\{ \sum_{i \in N} \ln s_i \ : \ s \in \mathcal{D}^* \right\}. \tag{7}$$

We conclude this paragraph by proving that the central solutions are the limits of the central paths of (P) and (D). The following theorem is due to McLinden [17]. The proof is new.

Theorem 2.7 *The central paths of (P) and (D) converge to the analytic centers of $\mathcal{P}^*$ and $\mathcal{D}^*$.*

Proof: In the proof of Theorem 2.5 we exhibited a decreasing sequence $\{\mu_k\}_{k=1}^{\infty}$ with the property that the associated sequences $\{x(\mu_k)\}$ and $\{s(\mu_k)\}$ on the central paths of (P) and (D) converge to limit points x^* in $\mathcal{P}^*$ and s^* in $\mathcal{D}^*$. Now let x^0 and s^0 be arbitrary points in $\mathcal{P}^*$ and $\mathcal{D}^*$. Then, using $(x^0 - x(\mu_k))^T(s^0 - s(\mu_k)) = 0$, we obtain $(x^0)^T s(\mu_k) + (s^0)^T x(\mu_k) = n\mu_k$, because $(s^0)^T x^0 = 0$. Using that $\sigma(x^0) \subseteq B$ and $\sigma(s^0) \subseteq N$, we have

$$\sum_{i \in B} x_i^0 s_i(\mu_k) + \sum_{i \in N} x_i(\mu_k) s_i^0 = n\mu_k.$$

Dividing both sides by μ_k and noting that $x_i(\mu_k)s_i(\mu_k) = \mu_k$, we obtain

$$\sum_{i \in B} \frac{x_i^0}{x_i(\mu_k)} + \sum_{i \in N} \frac{s_i^0}{s_i(\mu_k)} = n.$$

Letting $k \to \infty$, this implies

$$\sum_{i \in B} \frac{x_i^0}{x_i^*} + \sum_{i \in N} \frac{s_i^0}{s_i^*} = n.$$

By choosing $s^0 = s^*$, we see that

$$\sum_{i \in B} \frac{x_i^0}{x_i^*} = |B|,$$

where $|B|$ is the size of B. Rewriting this, we obtain

$$e_B^T(X_B^*)^{-1}(x_B^0 - x_B^*) = 0, \quad \forall x^0 \in \mathcal{P}^*.$$

Now, using Theorem 2.2, we observe that this is precisely the necessary and sufficient condition for x_B^* to be an optimal solution of the optimization problem (6). Therefore, x_B^* must be the central solution of (P). Until now, x_B^* was defined as an arbitrary limit point of the central path of (P). But the above argument makes clear that there is only one such limit point, namely the central solution of (P). This proves the theorem for the primal problem. A similar proof can be given for the dual problem . $\square$

Without going further into it we mention that every μ–center can be obtained as an analytic center, namely of the 'section' of the feasible region containing all feasible points having the same objective value as the μ–center.

2.2   Relaxing the interior point assumption

In this paragraph we remove the interior point Assumption 2.1. We present a new proof of the classical LP duality theorem, including the cases where one or both of the feasible regions of the problems (P) and (D) have empty (relative) interiors. This includes the cases that (P) and/or (D) is infeasible or unbounded. We also obtain the Goldman–Tucker theorem for this general case.

Choose arbitrary positive vectors $x^0, s^0 \in \mathbb{R}^n$ and an arbitrary vector $y^0 \in \mathbb{R}^m$. Also choose "large" numbers M_p and M_d such that

Assumption 2.2 $M_p > (b - Ax^0)^T y^0, \quad M_d > (A^T y^0 + s^0 - c)^T x^0.$

For the moment we shall call any such parameter pair (M_p, M_d) *feasible*.

We proceed by embedding the original problem pair $(P) - (D)$ in a pair $(\tilde{P}) - (\tilde{D})$ of slightly larger linear programming problems as follows.

$(\tilde{P})$ min $c^T x + M_p x_{n+1}$

   s.t. $Ax + (b - Ax^0)x_{n+1} = b$

      $(A^T y^0 + s^0 - c)^T x + x_{n+2} = M_d$

      $x \geq 0, x_{n+1} \geq 0, x_{n+2} \geq 0,$

$(\tilde{D})$ max $b^T y + M_d y_{m+1}$

   s.t. $A^T y + (A^T y^0 + s^0 - c)y_{m+1} + s = c$

      $(b - Ax^0)^T y + s_{n+1} = M_p$

      $y_{m+1} + s_{n+2} = 0$

      $s \geq 0, s_{n+1} \geq 0, s_{n+2} \geq 0.$

Let us write $\tilde{x} = (x, x_{n+1}, x_{n+2})$, $\tilde{s} = (s, s_{n+1}, s_{n+2})$ with $x, s \in \mathbb{R}^n$ and $\tilde{y} = (y, y_{m+1})$, $y \in \mathbb{R}^m$. Also define the $(m+1) \times (n+2)$ matrix $\tilde{A}$ and the vectors $\tilde{b} \in \mathbb{R}^{m+1}$, $\tilde{c} \in \mathbb{R}^{n+2}$,

$$\tilde{b} = \begin{pmatrix} b \\ M_d \end{pmatrix}, \quad \tilde{c} = \begin{pmatrix} c \\ M_p \\ 0 \end{pmatrix}, \quad \tilde{A} = \begin{pmatrix} A & b - Ax^0 & 0 \\ (A^T y^0 + s^0 - c)^T & 0 & 1 \end{pmatrix}.$$

We can then write $(\tilde{P})$ and $(\tilde{D})$ in a more compact form:

$(\tilde{P})$ min $\left\{ \tilde{c}^T \tilde{x} : \tilde{A}\tilde{x} = \tilde{b}, \tilde{x} \geq 0 \right\}$,

$(\tilde{D})$ max $\left\{ \tilde{b}^T \tilde{y} : \tilde{A}^T \tilde{y} + \tilde{s} = \tilde{c}, \tilde{s} \geq 0 \right\}$.

It is easy to see that the pair $(\tilde{P}) - (\tilde{D})$ satisfies Assumption 2.1 by considering the point $(\tilde{x}^0, \tilde{y}^0, \tilde{s}^0)$:

$$
\begin{aligned}
\tilde{x}^0 &= (x^0, 1, x^0_{n+2}) \in \mathbb{R}^{n+2}, \\
\tilde{s}^0 &= (s^0, s^0_{n+1}, 1) \in \mathbb{R}^{n+2}, \\
\tilde{y}^0 &= (y^0, -1) \in \mathbb{R}^{m+1},
\end{aligned}
$$

with $x^0_{n+2} = M_d - (A^T y^0 + s^0 - c)^T x^0$, and $s^0_{n+1} = M_p - (b - Ax^0)^T y^0$. We have $\tilde{A}\tilde{x}^0 = \tilde{b}$ and $\tilde{A}^T \tilde{y}^0 + \tilde{s}^0 = \tilde{c}$. Since $\tilde{x}^0$ and $\tilde{s}^0$ are positive (because of Assumption 2.2), Assumption 2.1 is satisfied.

The augmented programming pair $(\tilde{P}) - (\tilde{D})$ has been used (e.g. in [15, 18, 20]) for obtaining an initial interior point from which to start an IPM. We use it here for a different, theoretical purpose.

We treat the pair $(\tilde{P}) - (\tilde{D})$ as being parametrized by (M_p, M_d). We denote by $\pi(M_p, M_d) = (\tilde{B}, \tilde{N})$ the optimal partition of the index set $\{1, \ldots, n, n+1, n+2\}$. The optimal faces of $(\tilde{P})$ and $(\tilde{D})$ are then given by

$$
\begin{aligned}
\tilde{P}^* &= \{\tilde{x} \geq 0 : \tilde{A}\tilde{x} = \tilde{b}, \ x_{\tilde{N}} = 0\}, \\
\tilde{D}^* &= \{(\tilde{y}, \tilde{s}) : \tilde{A}^T \tilde{y} + \tilde{s} = \tilde{c}, \ \tilde{y} \in \mathbb{R}^{m+1}, \ \tilde{s} \geq 0, \ s_{\tilde{B}} = 0\},
\end{aligned}
$$

respectively. The following simple result is crucial.

Lemma 2.4 *Assume* (M^i_p, M^i_d), $i = 1, 2$, *are two feasible pairs such that* $\pi := (\tilde{B}, \tilde{N}) = \pi(M^i_p, M^i_d)$. *Let*

$$
\begin{aligned}
M_p &= (1 - \lambda_p) M^1_p + \lambda_p M^2_p, \\
M_d &= (1 - \lambda_d) M^1_d + \lambda_d M^2_d,
\end{aligned}
$$

respectively, where $0 \leq \lambda_p, \lambda_d \leq 1$. *Then,* $\pi(M_p, M_d) = \pi$.

Proof: Let $(\tilde{x}^i, \tilde{y}^i, \tilde{s}^i)$ be the central solution for the pair $(\tilde{P}) - (\tilde{D})$ with parameter pair (M^i_p, M^i_d), $i = 1, 2$. We have $(\sigma(\tilde{x}^i), \sigma(\tilde{s}^i)) = (\tilde{B}, \tilde{N})$. Consider

$$
\begin{aligned}
\tilde{x} &= (1 - \lambda_d) \tilde{x}^1 + \lambda_d \tilde{x}^2, \\
\tilde{y} &= (1 - \lambda_p) \tilde{y}^1 + \lambda_p \tilde{y}^2, \\
\tilde{s} &= (1 - \lambda_p) \tilde{s}^1 + \lambda_p \tilde{s}^2.
\end{aligned}
$$

One easily verifies that $(\tilde{x}, \tilde{y}, \tilde{s})$ is feasible for $(\tilde{P}) - (\tilde{D})$ with parameter pair (M_p, M_d) and, moreover, that $(\sigma(\tilde{x}), \sigma(\tilde{s})) = (\tilde{B}, \tilde{N})$. Therefore, $(\tilde{x}, \tilde{y}, \tilde{s})$ is a strictly complementary optimal solution for $(\tilde{P})-(\tilde{D})$ with parameter pair (M_p, M_d). Now Theorem 2.6 implies that $(\tilde{B}, \tilde{N})$ is the optimal partition in this case. $\square$

Let us consider any fixed partition $\pi = (\tilde{B}, \tilde{N})$ of the index set $\{1, \ldots, n+2\}$. We define the *optimality set* of π as the set of all feasible parameter pairs (M_p, M_d) such that $\pi(M_p, M_d) = \pi$. A direct consequence of Lemma 2.4 is that a nonempty optimality set is "rectangular" in the sense that its closure is a closed rectangle (which may be unbounded).

We draw an important conclusion from this. Since the number of partitions of the index set $\{1, \ldots, n+2\}$ is finite, the number of optimality sets is finite. These sets cover all feasible pairs (M_p, M_d). Hence, one of the optimality sets must contain all "large" feasible pairs (M_p, M_d). This implies the following result.

Theorem 2.8 *There exists a feasible pair (M_p^*, M_d^*) such that $\pi(M_p, M_d) = \pi(M_p^*, M_d^*)$ for all feasible pairs (M_p, M_d) with $M_p \geq M_p^*$ and $M_d \geq M_d^*$.*
$\square$

In the remainder of this paragraph the pair (M_p^*, M_d^*) will be as described in Theorem 2.8. Also, $\tilde{\pi}^* := (\tilde{B}^*, \tilde{N}^*)$ will denote the optimal partition for this pair, and $(\tilde{x}^*, \tilde{y}^*, \tilde{z}^*)$ the central solution. For all pairs (M_p, M_d) with $M_p \geq M_p^*$ and $M_d \geq M_d^*$ the optimal partition is given by $\tilde{\pi}^*$. Let $z^*(M_p, M_d)$ denote the common optimal value for the augmented problems. The next result shows that $z^*(M_p, M_d)$ is a bilinear function within the optimality set of $\tilde{\pi}^*$.

Lemma 2.5 *Let $M_p \geq M_p^*$ and $M_d \geq M_d^*$. Then*

$$z^*(M_p, M_d^*) = z^*(M_p^*, M_d^*) + \tilde{x}_{n+1}^*(M_p - M_p^*),$$
$$z^*(M_p^*, M_d) = z^*(M_p^*, M_d^*) - \tilde{s}_{n+2}^*(M_d - M_d^*).$$

Proof: Note that the feasible regions of $(\tilde{P})$ are the same for the pair (M_p, M_d^*) and the pair (M_p^*, M_d^*). Since the optimal partition is also equal for both pairs, both pairs will have the same primal central solution $\tilde{x}^*$. From the definition of $(\tilde{P})$ it now follows that $z^*(M_p, M_d^*)$ will depend linearly on M_p. This implies the first equality in the lemma. The proof of the second equality is similar. $\square$

The following result implies the LP duality theorem for an arbitrary linear program as well as the Goldman–Tucker theorem. It is the main result of this section. We use the following notation:

$$\tilde{x}^* \;=\; (x^*, x^*_{n+1}, x^*_{n+2}),\; x^* \in \mathbb{R}^n,$$
$$\tilde{s}^* \;=\; (s^*, s^*_{n+1}, s^*_{n+2}),\; s^* \in \mathbb{R}^n,$$
$$\tilde{y}^* \;=\; (y^*, y_{m+1}),\; y^* \in \mathbb{R}^m.$$

Theorem 2.9 *Let* (M^*_p, M^*_d) *and* $(\tilde{x}^*, \tilde{y}^*, \tilde{s}^*)$ *be as defined above. Then,*

(i) *if* $x^*_{n+1} > 0$ *and* $s^*_{n+2} > 0$, *then both* (P) *and* (D) *are infeasible;*

(ii) *if* $x^*_{n+1} > 0$ *and* $s^*_{n+2} = 0$, *then* (P) *is infeasible and* (D) *unbounded;*

(iii) *if* $x^*_{n+1} = 0$ *and* $s^*_{n+2} > 0$, *then* (P) *is unbounded and* (D) *infeasible;*

(iv) *if* $x^*_{n+1} = 0$ *and* $s^*_{n+2} = 0$, *then* (x^*, s^*) *is a strictly complementary optimal pair for* (P) *and* (D).

Proof: We claim that $x^*_{n+1} > 0$ if and only if (P) is infeasible. It is clear that if (P) is infeasible, then $x^*_{n+1} > 0$. To prove the converse, assume that $\bar{x}$ is feasible for (P). Choose $\bar{x}_{n+2} > 0$ such that $M^1_d :=$ $(A^T y^0 + s^0 - c)^T \bar{x} + \bar{x}_{n+2} \geq M^*_d$. Then $\tilde{x}^0 := (\bar{x}, 0, \bar{x}_{n+2})$ will be feasible for $(\tilde{P})$ with parameter M^1_d. Let $\tilde{x}^1$ be the central solution of $(\tilde{P})$ with the same parameter M^1_d. Then, using Lemma 2.5, we may write

$$z^*(M_p, M^1_d) \;=\; z^*(M^*_p, M^1_d) + x^1_{n+1}(M_p - M^*_p),$$
$$\;=\; z^*(M^*_p, M^*_d) - s^*_{n+2}(M^1_d - M^*_d) + x^1_{n+1}(M_p - M^*_p).$$

The objective value of $(\tilde{P})$ at $\tilde{x}^0$, which is $c^T \bar{x}$, is an upper bound for $z^*(M_p, M^1_d)$, for every value of $M_p \geq M^*_p$. However, due to Theorem 2.8 we have $\pi(M_p, M^1_d) = \pi(M_p, M^*_d)$, and thus $x^1_{n+1} > 0$. Therefore we have $\lim_{M_p \to \infty} z^*(M_p, M^1_d) = \infty$. This contradiction proves the claim. In the same way, one can show that $s^*_{n+2} > 0$ if and only if (D) is infeasible. So the proof of (i) is immediate.

We now prove (ii). From the above paragraph it is clear that (P) is infeasible and (D) is feasible in this case. Moreover, $\lim_{M_p \to \infty} z^*(M_p, M^*_d) = \infty$. Since $y^*_{m+1} = -s^*_{n+2} = 0$, we have $\tilde{y}^* = (y^*, 0)$. Hence, y^* is feasible for (D) and $\tilde{b}^T \tilde{y}^* = b^T y^* = z^*(M_p, M^*_d)$. Since $\lim_{M_p \to \infty} z^*(M_p, M^*_d) = \infty$ we conclude that (D) must be unbounded. The proof of (iii) is similar.

It remains to prove (iv). Since $x_{n+1}^* = 0$ and $s_{n+2}^* = 0$, x^* is feasible for (P) and y^* is feasible for (D). Furthermore, $c^T x^* = \tilde{c}^T \tilde{x}^* = \tilde{z}(M_p, M_d) = \tilde{b}^T \tilde{y}^* = b^T y^*$. Hence, we have strong duality for the problems (P) and (D). Moreover, since the pair $(\tilde{x}^*, \tilde{s}^*)$ is strictly complementary, the pair (x^*, s^*) is also strictly complementary. $\hfill\square$

Theorem 2.9 makes clear that the notion of *optimal partition* can be extended to the case that Assumption 2.1 is not satisfied, namely by taking the restriction of the partition $\tilde{\pi}^* := (\tilde{B}^*, \tilde{N}^*)$ defined above to the index set $\{1, \ldots, n\}$.

3   A polynomial–time path–following algorithm

3.1   Introduction

Since Karmarkar [14] published his method for the solution of the linear programming problem, world–wide research has resulted in a wide variety of interior point methods for linear programming. The various methods can be divided in four categories:

1) Projective methods, whose more typical representants are Karmarkar [14] and the extensions of Anstreicher [1], de Ghellink and Vial [5] and Todd and Burrell [29].

2) Pure affine scaling methods, originally due to Dikin [4], and subsequently rediscovered by Barnes [3] and Vanderbei et al. [32].

3) Path-following methods, initially due to Renegar [21] and extended by Vaidya [31], Gonzaga [10], Goldfarb and Liu [8], Monteiro and Adler [20], Kojima, Mizuno and Yoshise [16], Roos [23] and Roos and Vial [24].

4) Potential-based affine scaling methods due to Ye [33] (see also Todd and Ye [30]), followed by Gonzaga [11], Freund [6] and Anstreicher and Bosch [2].

For each of the methods in the first, second and fourth category, a polynomial–time bound has been derived. Despite the fact that these methods at first sight seem to be quite different, they have a common feature, namely that they use the central path of the problem as a guide to the central solution of the problem. This is not the case with the methods in the second category. Maybe this is the reason that for the methods in the second category only global convergence results have been proved.

In this paper we shall present a method which belongs to the third category. The key ingredients in any path-following method are the following:

a) a parametrization of the central path (in our case μ is the parameter);

b) an iterative process which brings the parameter to the value which is associated with the central solution of the problem ($\mu = 0$ in our case);

c) a concept of neighbourhood of a point on the central path; and finally

d) a step in the space of variables of the problem, which takes a point in the neighborhood of a center to the neighborhood of a new center corresponding to the new value of the parameter in b).

Gonzaga [10] was the first who devised a method based on the use of the logarithmic barrier function. Later on Roos and Vial [24] did the same, but by using a different distance measure they were able to simplify the analysis of the method considerably. In both cases the search direction is simply the (projected) Newton step associated with the minimization of the logarithmic barrier function.

Below we will present the algorithm and the complexity analysis of Roos and Vial. In the next paragraph the search direction and the distance measure are introduced. It will turn out that, when the current iterate is at a 'small enough' distance from a center, the process of repeatedly applying pure projected Newton steps converges quadratically to this center. This property plays an important role in the convergence analysis, and allows great simplification in the proofs to yield a simple and elegant presentation of the method.

Although the derived iteration bound is the best achieved so far by interior point methods, a direct implementation of a path-following method may yield disappointing results. The number of reductions of the parameter value required to reach optimality may be very large. Since computing a projected Newton step is a major computational task, the procedure may require a prohibitive amount of computing time. In the concluding paragraph of this section we suggest a practical implementation which departs from what the theory recommends but which might turn out to be more efficient.

It will be assumed throughout that the feasible regions of both (P) and (D) have nonempty interior.

3.2   Search direction and distance measure

Suppose that we are given x such that $Ax = b$ and $x > 0$, i.e. x is strictly feasible for (P). The search direction in our algorithm is simply the projected Newton direction of the logarithmic barrier function, which is given by

$$p(x, \mu) := -X P_{AX} \left(\frac{Xc}{\mu} - e \right),$$

where P_{AX} denotes the orthogonal projection on the null space of the matrix AX. This search direction has a nice interpretation which, strictly spoken, is not necessary for the analysis of our method, but which may yield a better understanding.

Lemma 3.1 *One has*

$$P_{AX} \left(\frac{Xc}{\mu} - e \right) = \frac{X s(x, \mu)}{\mu} - e,$$

where

$$s(x, \mu) := \operatorname{argmin}_s \left\{ \left\| \frac{Xs}{\mu} - e \right\| : A^T y + s = c, \quad y \in \mathbb{R}^m \right\}. \qquad \Box$$

Here $\|.\|$ denotes the 2–norm. The proof of Lemma 3.1 is omitted; it is straightforward from writing down first–order optimality conditions.

The 2–norm of $P_{AX} \left(\frac{Xc}{\mu} - e \right)$ will be used as a measure for the distance of the given point x to the μ–center $x(\mu)$. It is denoted as $\delta(x, \mu)$. So, from Lemma 3.1 we have

$$\delta(x, \mu) := \left\| \frac{X s(x, \mu)}{\mu} - e \right\| = \| X^{-1} p(x, \mu) \|.$$

It may be worthwhile to indicate that $\delta(x(\mu), \mu) = 0$ and $s(x(\mu), \mu) = s(\mu)$.

The point which results by taking the projected Newton step at x will be denoted by x^*, and is given by

$$x^* := x + p(x, \mu) = x - X \left(\frac{X s(x, \mu)}{\mu} - e \right) = 2x - \frac{X^2 s(x, \mu)}{\mu}.$$

We now are able to state and prove some elementary results which are crucial for our approach.

Theorem 3.1 *If $\delta(x, \mu) < 1$ then x^* is a strictly feasible point for (P). Moreover, $\delta(x^*, \mu) \leq \delta(x, \mu)^2$.*

Proof: We use the short hand notation $p := p(x, \mu)$. One has $x^* = x + p = X(e + X^{-1}p) > 0$, because $\|X^{-1}p\| = \delta(x, \mu) < 1$. Moreover, $Ax^* = A(x + p) = b$, because p lies in the null space of A. This proves the first part of the theorem.

Note that the definition of $s(x^*, \mu)$ implies the following:

$$\delta(x^*, \mu) = \left\| \frac{X^* s(x^*, \mu)}{\mu} - e \right\| \le \left\| \frac{X^* s(x, \mu)}{\mu} - e \right\|.$$

Due to Lemma 3.1 we have

$$s(x, \mu) = \mu X^{-1}(e - X^{-1}p).$$

Also using $X^* = X + P$ we now may write

$$\frac{X^* s(x, \mu)}{\mu} - e = (X + P)X^{-1}(e - X^{-1}p) - e = X^{-2}Pp.$$

Substitution gives

$$\delta(x^*, \mu) \le \|X^{-2}Pp\| \le \|X^{-1}p\|^2 = \delta(x, \mu)^2,$$

which was to be shown.   □

The importance of Theorem 3.1 is clear: if we repeatedly replace x by x^*, we obtain a sequence of points which quadratically converges to the μ−center $x(\mu)$.

Theorem 3.2 Let $\mu^* := (1 - \theta)\mu$, with $0 < \theta < 1$. Then $\delta(x, \mu^*) \le \frac{1}{1-\theta}(\delta(x, \mu) + \theta\sqrt{n})$.

Proof: Using the definition of $s(x, \mu^*)$ we may write

$$\delta(x, \mu^*) = \left\| \frac{X s(x, \mu^*)}{\mu^*} - e \right\| \le \left\| \frac{X s(x, \mu)}{\mu^*} - e \right\|.$$

Denoting $t := \frac{X s(x, \mu)}{\mu}$, we obtain

$$\delta(x, \mu^*) \le \left\| \frac{t}{1 - \theta} - e \right\| \le \frac{\|t - e\| + \theta\|e\|}{1 - \theta}.$$

Since $\|t - e\| = \delta(x, \mu)$ and $\|e\| = \sqrt{n}$, this implies the theorem.   □

Theorem 3.3 Let $\delta(x, \mu) \le 1/2$. If $\theta = 1/(6\sqrt{n})$ and x^* and μ^* are as defined before, then $\delta(x^*, \mu^*) \le 1/2$.

Proof: Using Theorem 3.2 and Theorem 3.1 successively we may write

$$
\begin{aligned}
\delta(x^*,\mu^*) &\leq \tfrac{1}{1-\theta}(\delta(x^*,\mu)+\theta\sqrt{n})\\
&\leq \tfrac{1}{1-\theta}(\delta(x,\mu)^2+\theta\sqrt{n})\\
&\leq \tfrac{1}{1-\theta}(\tfrac{1}{4}+\tfrac{1}{6})\\
&= \tfrac{5}{12(1-\theta)}\\
&\leq \tfrac{1}{2},
\end{aligned}
$$

and hence the result follows.   □

Now suppose that we are given x and μ such $\delta(x,\mu)\leq 1/2$. It may be mentioned that every linear programming problem is polynomially equivalent to a linear programming problem in standard form for which such an x and μ are known. In fact, this can be accomplished by using the embedding of Section 2.2 with M_p and M_d suitably choosen. See, e.g., Renegar [21] or Monteiro et al. [20].

It is obvious from Theorem 3.3 that we can construct a sequence of pairs (x_i,μ_i), $i=0,1,2,\ldots$, such that for each i the point x is strictly feasible for (P) and $\mu_i > 0$, whereas $\delta(x_i,\mu_i)\leq 1/2$, and $\mu_i \to 0$ if $i \to \infty$, namely by taking $x_0 := x$, $\mu_0 := \mu$, and $x_{i+1} := x_i^*$, $\mu_{i+1} := (1 - 1/(6\sqrt{n}))\mu_i$ for $i \geq 0$. In this connection the following result is important.

Theorem 3.4 *If $\delta(x,\mu)\leq 1$ then $y(x,\mu)$ is dual feasible. Moreover,*

$$
\mu(n-\delta(x,\mu)\sqrt{n}) \leq c^T x - b^T y(x,\mu) \leq \mu(n+\delta(x,\mu)\sqrt{n}).
$$

Proof: Since $\delta(x,\mu) = \|\frac{Xs(x,\mu)}{\mu} - e\|$ it follows from $\delta(x,\mu)\leq 1$ that $Xs(x,\mu) \geq 0$, whence $s(x,\mu) \geq 0$. Consequently, $A^T y(x,\mu) \leq c$, which means that $y(x,\mu)$ is dual feasible. This proves the first statement in the theorem. Using $Ax = b$ we now may write

$$
\begin{aligned}
c^T x - b^T y(x,\mu) &= c^T x - x^T A^T y(x,\mu)\\
&= c^T x - x^T(c - s(x,\mu))\\
&= x^T s(x,\mu).
\end{aligned}
$$

Furthermore, application of the Cauchy-Schwarz inequality gives

$$
\delta(x,\mu)\sqrt{n} = \|\frac{Xs(x,\mu)}{\mu} - e\|\|e\| \geq \|\frac{x^T s(x,\mu)}{\mu} - n\|,
$$

which implies that

$$
n - \delta(x,\mu)\sqrt{n} \leq \frac{x^T s(x,\mu)}{\mu} \leq n + \delta(x,\mu)\sqrt{n}.
$$

Hence the theorem follows. $\square$

The above results suggest the following algorithm for the solution of (P).

Primal Algorithm

Input:
A pair (x_0, μ_0) such that x_0 is strictly feasible, $\mu_0 > 0$ and $\delta(x_0, \mu_0) \leq 1/2$.
An accuracy parameter $q \in \mathbb{N}$.

begin
$\quad \theta := 1/(6\sqrt{n})$; $x := x_0$; $\mu := \mu_0$;
$\quad$ **while** $n\mu > e^{-q}$ **do**
$\quad$ **begin**
$\quad\quad x := x + p(x, \mu)$;
$\quad\quad \mu := (1 - \theta)\mu$;
$\quad$ **end**
end.

Theorem 3.5 *Let $q_0 := -\ln(n\mu_0)$. Then the Primal Algorithm stops after at most $6(q - q_0)\sqrt{n}$ steps. The last generated point x and $y(x, \mu)$ are both strictly feasible, whereas $c^T x - b^T y(y, \mu) \leq \frac{3}{2} e^{-q}$.*

Proof: After each iteration of the algorithm x will be strictly feasible, and $\mu > 0$, whereas $\delta(x, \mu) \leq 1/2$, due to Theorem 2.3. After the k-th iteration we will have $\mu = (1 - \theta)^k \mu_0$. The algorithm stops if k is such that

$$(1 - \theta)^k e^{-q_0} < e^{-q}.$$

Taking logarithms, this inequality reduces to

$$-k \ln(1 - \theta) > q - q_0.$$

Since $-\ln(1 - \theta) > \theta$, this will certainly hold if

$$k\theta > q - q_0,$$

which is equivalent to

$$k > 6(q - q_0)\sqrt{n}.$$

This proves the first statement in the theorem. Now let x be the last generated point. Then Theorem 2.4 implies that $y(x, \mu)$ is dual feasible and

$$
\begin{aligned}
c^T x - b^T y &\leq \mu(n + \delta(x, \mu)\sqrt{n}) \\
&\leq e^{-q}(1 + \frac{\delta(x, \mu)}{\sqrt{n}}) \\
&\leq \frac{3}{2} e^{-q}.
\end{aligned}
$$

This completes the proof. $\square$

3.3 Algorithm for the dual problem

In this paragraph we will briefly describe how a polynomial algorithm for the dual problem (D) can be obtained by using a quite similar approach as described before for the primal problem (P). We shall assume that we are given y and s such that $A^T y + s = c$, with $s > 0$. So y is strictly feasible for (D). Defining

$$x(y, \mu) := \text{argmin}_x \left\{ \left\| \frac{Sx}{\mu} - e \right\| : Ax = b, \quad x \in \mathbb{R}^n \right\},$$

we will have

$$S \left(\frac{Sx(y, \mu)}{\mu} - e \right) = A^T p$$

for some unique $p \in \mathbb{R}^m$. One may easily check that the vector p is precisely the Newton step at y for the logarithmic barrier function associated with (D). It will be denoted as $p(y, \mu)$. We will use the point $y^* := y + p(y, \mu)$ as the next iterate in our algorithm. Defining $s^* := c - A^T y^*$, and $p(s, \mu) = -A^T p(y, \mu)$ one has

$$s^* = s + p(s, \mu) = s - S \left(\frac{Sx(y, \mu)}{\mu} - e \right) = 2s - \frac{S^2 x(y, \mu)}{\mu}.$$

The 2-norm of the vector $\frac{Sx(y, \mu)}{\mu} - e$ is a measure for the distance from the point y to the point $y(\mu)$ on the central path. So, denoting this distance measure as $\delta(y, \mu)$, we have

$$\delta(y, \mu) := \left\| \frac{Sx(y, \mu)}{\mu} - e \right\| = \| S^{-1} p(s, \mu) \|.$$

We now can state the following results, whose proofs are omitted here, since they are completely similar to the proofs of the corresponding results for the primal case.

Theorem 3.6 If $\delta(y, \mu) < 1$, then $s^* > 0$, i.e. y^* is strictly feasible. Moreover, $\delta(y^*, \mu) \le \delta(y, \mu)^2$. □

Theorem 3.7 Let $\mu^* := (1 - \theta)\mu$, with $0 < \theta < 1$. Then $\delta(y, \mu^*) \le \frac{1}{1-\theta}(\delta(y, \mu) + \theta \sqrt{n})$. □

Theorem 3.8 Let $\delta(y, \mu) \le 1/2$. If $\theta = 1/(6\sqrt{n})$ and y^* and μ^* are as before, then $\delta(y^*, \mu^*) \le 1/2$. □

Theorem 3.9 *If $\delta(y,\mu) \leq 1$ then $x(y,\mu)$ is primal feasible. Moreover,*

$$\mu(n - \delta(y,\mu)\sqrt{n}) \leq c^T x - b^T y \leq \mu(n + \delta(y,\mu)\sqrt{n}). \qquad \square$$

Just as in the primal case the above results suggest a method for the solution of problem (D).

Dual Algorithm

Input:
A pair (y_0, μ_0) such that y_0 is strictly feasible, $\mu_0 > 0$ and $\delta(y_0, \mu_0) \leq 1/2$. An accuracy parameter $q \in \mathbb{N}$.

begin
    $\theta := 1/(6\sqrt{n})$; $y := y_0$; $\mu := \mu_0$;
    while $n\mu > e^{-q}$ **do**
    begin
        $y := y + p(y,\mu)$;
        $\mu := (1 - \theta)\mu$;
    end
end.

Finally, copying the proof for the primal case one gets the polynomial–time convergence property of this algorithm.

Theorem 3.10 *Let $q_0 := -\ln(n\mu_0)$. Then the Dual Algorithm stops after at most $6(q - q_0)\sqrt{n}$ steps. The last generated point y and $x(y,\mu)$ are both strictly feasible, whereas $c^T x(y,\mu) - b^T y \leq \frac{3}{2}e^{-q}$.*

3.4   Implementation strategies

The polynomial convergence property of the algorithms presented in the preceding paragraphs stems from the reduction of the parameter μ at each iteration by a factor $1 - \theta = 1 - 1/(6\sqrt{n})$. For practical problems, this factor is close to 1. Consequently, the number of iterations which will be necessary to achieve the target value $\mu \leq e^{-q}/n$, will be very large. In view of the high computational cost of each individual iteration, a straightforward implementation of the algorithm may turn out to be inefficient.
In this paragraph we propose three possible remedies. Thereby we restrict ourselves to the primal algorithm; similar arguments apply to the dual algorithm.

1. At no cost we can make at each iteration the reduction of μ as large as possible by solving

$$\min \left\{ \mu : \left\| P_{AX} \left(\frac{Xc}{\mu} - e \right) \right\| \leq 0.5 \right\}.$$

This strategy is discussed by Ye [33] and also by Mizuno and Todd [19], and can be implemented quite efficiently.

2. We can even go further. At each iteration solve

$$\min \left\{ \mu : \left\| P_{AX} \left(\frac{Xc}{\mu} - e \right) \right\| \leq \theta \right\},$$

and do a few Newton steps, with μ fixed, until a point x is obtained such that $\delta(x, \mu) \leq 0.5$ again. For instance, with $\theta = 0.95$, Theorem 3.1 implies that at most four Newton steps are required to achieve the result.

3. One may be tempted to enforce much larger reductions in μ, by taking $\theta = 0.5$, say. Unfortunately it is then no longer possible to guarantee that the next iterate will remain in the vicinity of the centre $x(\mu)$, as measured by the distance $\delta(x, \mu)$. Moreover, if $\delta(x, \mu) > 1$, the step may well bring the iterate outside the positive orthant. However, recall that the step in the variable space is just the Newton step associated with the logarithmic barrier function. This function is well–behaved. It is strongly convex and it tends towards $+\infty$ near the boundary of the positive orthant. Thus, it is possible to implement a safeguarded line search along the Newtonian direction to minimize the barrier function. This will prevent us from bringing the next iterate outside the positive orthant. Moreover, if x is close enough to the μ–center, $\delta(x, \mu) < 1$, then, by Theorem 2.1, Newton's method (with no update of μ) will quadratically converge to $x(\mu)$. So, one can think of an algorithm which alternates significant reductions of the parameter μ, say by putting $\mu := \mu/2$, and one or several safeguarded Newtonian steps to bring the iterate in the vicinity of $x(\mu)$. The closeness factor can be taken arbitrarily, for example $\delta(x, \mu) \leq 0.5$. This proposal is very much in the spirit of the classical barrier method, as presented in Gill et al.. [7] and is worked out in Roos and Vial [25].

4   Parametric analysis

Let $1 \leq j \leq n$ and $1 \leq i \leq m$. Denoting the objective coefficient c_j as γ and the right hand side coefficient b_i as β, we investigate in this section the dependency of the optimal solutions of (P) and (D) on the pair (β, γ). More

precisely, we shall be interested in the behaviour of the central solutions
of (P) and (D) as a function of the pair (β,γ). It is clear that the central
solutions exist only if β and γ are such that the feasible regions of (P) and
(D) contain positive vectors. If this is the case we shall call (β,γ) a *feasible
pair*. We shall denote the corresponding optimal partition as $\pi(\beta,\gamma)$, and
the optimal value as $z(\beta,\gamma)$.

Now let us consider any fixed ordered partition $\pi = (B,N)$ of the index
set $\{1,\ldots,n\}$. Then we may consider the set of all pairs (β,γ) such that
$\pi(\beta,\gamma) = \pi$. We shall call this the *optimality set* of π, denoted as $G(\pi)$. So

$$G(\pi) := \{(\beta,\gamma) \ : \ \pi(\beta,\gamma) = \pi\}.$$

Clearly, an optimality set may be empty. On the other hand, every pair
(β,γ) belongs to a unique optimality set. Therefore, the optimality sets
form a partition of the set of all feasible pairs. At first sight it might not
be clear that the optimality sets are connected. But our next result not
only implies that the optimality sets are connected, but also that they are
rectangular.

Lemma 4.1 *Let (β_1,γ_1) and (β_2,γ_2) denote feasible pairs which determine
the same partition of the index set $\{1,\ldots,n\}$. This (ordered) partition will
be denoted as $\pi = (B,N)$. If β is any convex combination of β_1 and β_2,
and γ is any convex combination of γ_1 and γ_2, then $\pi(\beta,\gamma) = \pi$.*

Proof: The proof strongly resembles the proof of Lemma 2.4 and is therefore omitted. $\square$

In the remaining part of this section we will further investigate the optimality sets. One of our aims is to determine the optimal partitions for the
optimality sets surrounding a given optimality set. Also, we are interested
in the behaviour of the optimal value $z(\beta,\gamma)$ within an optimality set and
on the boundary of optimality sets. Before proceeding we introduce some
more notation.

Given an ordered partition $\pi = (B,N)$ we define

$$\begin{aligned}
\mathcal{P}(\pi) &= \{x \ : \ Ax = b, \ x_B \geq 0, \ x_N = 0\} \\
\mathcal{D}(\pi) &= \{(y,s) \ : \ A^Ty + s = c, \ s_N \geq 0, \ s_B = 0\}.
\end{aligned}$$

Note that if the interiors of $\mathcal{P}(\pi)$ and $\mathcal{D}(\pi)$ are not empty then these are the
optimal sets for (P) and (D), respectively, and π is the optimal partition for

(P) and (D). This follows by noting that the duality gap is zero whenever $x \in \mathcal{P}(\pi)$ and $(y, s) \in \mathcal{D}(\pi)$. As a consequence, given (P), and (D), there is exactly one partition π such that the interiors of both $\mathcal{P}(\pi)$ and $\mathcal{D}(\pi)$ are nonempty, and this π is the optimal partition for (P) and (D). The analytic centers of $\mathcal{P}(\pi)$ and $\mathcal{D}(\pi)$ will be denoted as $x(\pi)$ and $y(\pi)$, and due to Theorem 2.7 these are the central solutions of (P) and (D).

We now turn to the study of the properties of $G(\pi)$. We will do this for β and γ separately. We will first consider the case that γ is fixed and investigate the effect of changes in β (right–hand–side–ranging). Later on we consider the case that β is fixed and γ is changed (cost–coefficient–ranging).

4.1   Right-hand–side–ranging

We will first consider the case that $c_i = \gamma$ is fixed and then investigate the effect of changes in $b_j = \beta$. With $\pi = \pi(\beta, \gamma)$, we will need the solutions of the linear programming problems

$$\min\{y_i \; : \; y \in \mathcal{D}(\pi)\}, \tag{8}$$

and

$$\max\{y_i \; : \; y \in \mathcal{D}(\pi)\}. \tag{9}$$

If they exist, the central solutions of these problems are denoted as $y^-(\pi)$ and $y^+(\pi)$ respectively, and the corresponding optimal partitions as π_d^- and π_d^+. We will use the notations $\pi = (B, N)$, $\pi_d^+ = (B^+, N^+)$, $\pi_d^- = (B^-, N^-)$. Furthermore, e_i will denote the ith unit vector.

Theorem 4.1 *Let (β, γ) be a feasible pair and $\pi := \pi(\beta, \gamma)$. Then there exist positive numbers δ^+ and δ^- such that*

$$\pi(\beta + \epsilon, \gamma) = \pi_d^+, \quad z(\beta + \epsilon, \gamma) = z(\beta, \gamma) + \epsilon y_i^+(\pi) \qquad (0 < \epsilon < \delta^+)$$
$$\pi(\beta - \epsilon, \gamma) = \pi_d^-, \quad z(\beta - \epsilon, \gamma) = z(\beta, \gamma) - \epsilon y_i^-(\pi) \qquad (0 < \epsilon < \delta^-).$$

Proof: By definition, $x_k(\pi)$ is positive if $k \in B$ and zero if $k \in N$. Changing $b_i = \beta$ to $\beta + \epsilon$, any feasible x must have the form $x = x(\pi) + \epsilon\xi$, where ξ is such that $A\xi = e_i$. Let $\xi(\pi)$ be the central solution of the problem

$$\min\{c^T\xi \; : \; A\xi = e_i, \; \xi_N \geq 0\}. \tag{10}$$

Note that $\xi_B(\pi)$ may have negative entries, but since $x_B(\pi)$ is positive there exists a positive δ^+ such that $x(\pi) + \delta^+\xi(\pi)$ is nonnegative. Hence, if

$0 < \epsilon < \delta^+$, then $x := x(\pi) + \epsilon\xi(\pi)$ is strictly feasible for the new situation and the objective value is given by

$$c^T x = c^T x(\pi) + \epsilon c^T \xi(\pi).$$

The dual problem of (10) is

$$\max\{y_i \ : \ A_B^T y = c_B, \ A_N^T y \le c_N\}, \tag{11}$$

which is exactly the problem (9). Hence, using strong duality, we have $c^T \xi(\pi) = y_i^+(\pi)$. So, also using that $c^T x(\pi) = z(\beta, \gamma)$, we may write

$$c^T x = z(\beta, \gamma) + \epsilon y_i^+(\pi). \tag{12}$$

By definition, π_d^+ is the optimal partition for (11), and hence also for (10). Also, $y^+(\pi)$, being the central solution of (11), is an interior point of $\mathcal{D}(\pi_d^+)$. We proceed by showing that x is an interior point of $\mathcal{P}(\pi_d^+)$. Note the obvious fact that $B \subseteq B^+$. For $k \in N$, it holds that $\xi_k(\pi) > 0$ if and only if $k \in B^+$, hence $x_k > 0$ if and only if $k \in B^+$. Thus it follows that $(x, y^+(\pi))$ is a strictly complementary optimal pair. Hence π_d^+ is the optimal partition in the new situation. As a consequence, the value of $z(\beta + \epsilon, \gamma)$ is given by (12). This proves the first part of the theorem. The proof of the second part goes in the same way, and is therefore omitted. □

Under a certain condition, the solutions of the problems (8) and (9) can be shown to be closely related to the solution of the original problem.

Theorem 4.2 *If e_i belongs to the column space of A_B then $\pi_d^- = \pi_d^+ = \pi$ and $y^-(\pi) = y^+(\pi) = y(\pi)$.*

Proof: If e_i belongs to the column space of A_B then there exists a vector u such that $A_B u = e_i$. Now let $y \in \mathcal{D}(\pi)$. Then we may write

$$y_i = y^T e_i = y^T A_B u = c_B^T u,$$

proving that y_i is constant on $\mathcal{D}(\pi)$. But then the central solution of the problems (8) and (9) is simply the analytic center $y(\pi)$ of $\mathcal{D}(\pi)$. So we have $y^-(\pi) = y^+(\pi) = y(\pi)$ and $\pi_d^- = \pi_d^+ = \pi$. Thus the theorem follows. □

A direct consequence of the last two results is

Corollary 4.1 *If e_i belongs to the column space of A_B then y_i is constant on $\mathcal{D}(\pi)$. Furthermore, the interval $G_\gamma(\pi) := \{\beta \ : \ \pi(\beta, \gamma) = \pi\}$ is open, and $z(\beta, \gamma)$ is linear in this interval, with slope $y_i(\pi)$.* □

The following theorem implies that the condition in Corollary 4.1 is necessary and sufficient for y_i to be constant on $\mathcal{D}(\pi)$.

Theorem 4.3 *If e_i does not belong to the column space of A_B then $\pi_d^- \neq \pi$ and $\pi_d^+ \neq \pi$. Furthermore, e_i belongs to the column space of A_{B^+} and to the column space of A_{B^-}. Moreover $y_i^-(\pi) < y_i(\pi) < y_i^+(\pi)$.*

Proof: If e_i does not belong to the column space of A_B then there exists no ξ such that $A\xi = e_i$ and $\xi_N = 0$. Hence the optimal partitions for the problems (8) and (9) must differ from π. Clearly, $y(\pi)$ is feasible for the problems (8) and (9). But since, in both cases, π differs from the optimal partition, $y(\pi)$ cannot be optimal. Hence we must have $y_i^-(\pi) < y_i(\pi) < y_i^+(\pi)$. Finally, if ξ^- is an optimal solution of (8) then $A_{B^-}\xi^- = e_i$, and hence e_i belongs to the column space of A_{B^-}. In the same way it follows that e_i belongs to the column space of A_{B^+}.   □

Corollary 4.2 *If e_i does not belong to the column space of A_B then the interval $G_\gamma(\pi) := \{\beta : \pi(\beta, \gamma) = \pi\}$ is a singleton. Moreover, $z(\beta, \gamma)$ has a breakpoint in β, with left derivative $y_i^-(\pi)$ and right derivative $y_i^+(\pi)$.* □

Summarizing the above results we may state

Theorem 4.4 *If γ is kept fixed, then $z(\beta, \gamma)$ is a piecewise linear convex function of β. The breakpoints occur where the optimal partition $\pi(\beta, \gamma)$ changes.* □

Example 4.1 In Figure 1 we have drawn $y_2(\pi)$ as a function of $\beta = b_2$ for a specific small example. The feasible region is in $\mathbb{R}^2$, and depicted left in the figure. The objective vector is $b = (1, \beta)$. In this example the breakpoints occur when b is a normal vector for one of the facets of the feasible region. In that case the central solution of the problem is the analytic center of this facet. Otherwise, in this example, the optimal solution is a vertex and this occurs if β is in the open interval between the neighbouring breaking values.

4.2   Cost–coefficient–ranging

Now we turn to the case that $\beta = b_i$ is fixed and $\gamma = c_j$ varies (Cost-coefficient-ranging). In this case we need to distinguish the cases $j \in N$ and $j \in B$. The first case is easy.

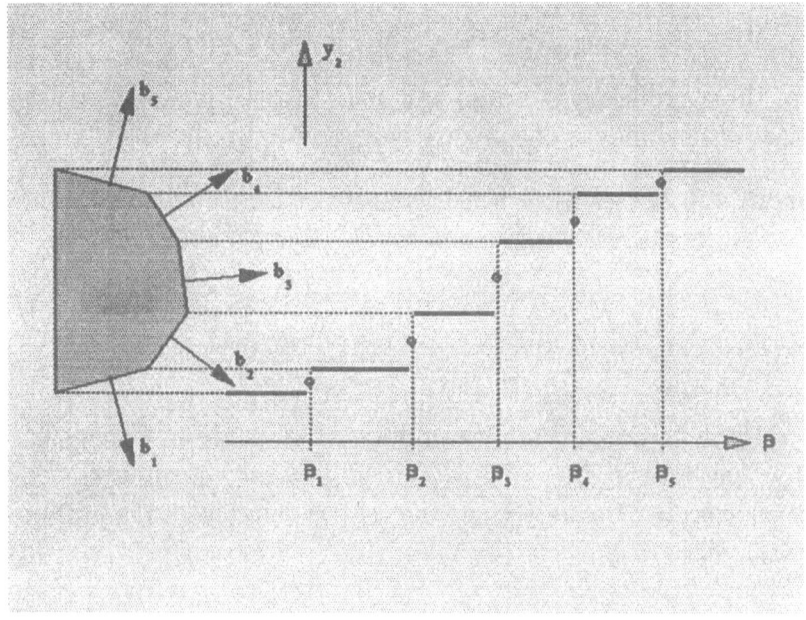

Figure 1: y_2 as a function of β.

Theorem 4.5 *Let* (β, γ) *be a feasible pair and* $\pi := \pi(\beta, \gamma) = (B, N)$. *If* $j \in N$ *then there exists a positive number* δ *such that for all* $\epsilon > -\delta$

$$\pi(\beta, \gamma + \epsilon) = \pi, \quad z(\beta, \gamma + \epsilon) = z(\beta, \gamma).$$

Proof: One has $A_B^T y(\pi) = c_B$ and $A_N^T y(\pi) < c_N$. If $j \in N$ then it is obvious that $y(\pi)$ is feasible for (D) for all values of c_j greater than $\gamma - \delta$, where $\delta < \gamma - a_j^T y(\pi)$. Hence the strictly complementary pair $(x(\pi), y(\pi))$ remains optimal in the new situation. Consequently, the optimal partition is π again. Since $x_j(\pi) = 0$ there will be no change in the optimal objective value and hence the theorem follows.   □

If $j \in B$ we will need the solutions of the linear programming problems

$$\min\{x_j \ : \ x \in \mathcal{P}(\pi)\} \tag{13}$$

and

$$\max\{x_j \ : \ x \in \mathcal{P}(\pi)\}. \tag{14}$$

It will turn out that the first problem plays a role if γ increases and the second if γ decreases. Therefore, if they exist, we denote the central solutions of these problems as $x^+(\pi)$ and $x^-(\pi)$ respectively and the optimal partitions of the index set $\{1,\ldots,n\}$ as π_p^+ and π_p^-; e_j is the jth unit vector.

Theorem 4.6 *Let (β,γ) be a feasible pair and $\pi := \pi(\beta,\gamma) = (B,N)$. If $j \in B$, then there exist positive numbers δ^+ and δ^- such that*

$$\pi(\beta,\gamma+\epsilon) = \pi_p^+, \quad z(\beta,\gamma+\epsilon) = z(\beta,\gamma) + \epsilon x_j^+(\pi) \quad (0 < \epsilon < \delta^+)$$
$$\pi(\beta,\gamma-\epsilon) = \pi_p^-, \quad z(\beta,\gamma-\epsilon) = z(\beta,\gamma) - \epsilon x_j^-(\pi) \quad (0 < \epsilon < \delta^-).$$

Proof: By definition, $A_B^T y(\pi) = c_B$ and $A_N^T y(\pi) < c_N$. Changing $c_j = \gamma$ to $\gamma + \epsilon$, any feasible y can be written as $y = y(\pi) + \epsilon\eta$, where η is such that $A_B^T \eta \leq (e_j)_B$. Define $\eta(\pi)$ as the central solution of the problem

$$\max\{b^T\eta : A_B^T\eta \leq (e_j)_B\}. \tag{15}$$

Since $A_N^T y(\pi) < c_N$, there exists a positive δ^+ such that $y(\pi) + \delta^+\eta(\pi)$ is feasible in the new situation. Hence, if $0 < \epsilon < \delta^+$ then $y := y(\pi) + \epsilon\eta(\pi)$ will be feasible as well in the new situation, and the objective value is given by

$$b^T y = b^T y(\pi) + \epsilon b^T \eta(\pi).$$

The dual problem of (15) can be written as

$$\min\{x_j : A_B x_B = b, \ x_B \geq 0, \ x_N = 0\}, \tag{16}$$

which is exactly the problem (13). Hence, using strong duality, we have $b^T\eta(\pi) = x_j^+(\pi)$. So, also using that $b^T y(\pi) = z(\beta,\gamma)$, we may write

$$b^T y = z(\beta,\gamma) + \epsilon x_j^+(\pi). \tag{17}$$

By definition, π_p^+ is the optimal partition for (16), and hence also for (15). Also $x^+(\pi)$, being the central solution of (16), is an interior point of $\mathcal{P}(\pi_p^+)$. We proceed by showing that y is an interior point of $\mathcal{D}(\pi_p^+)$. Since $N \subseteq N^+$, one has $B^+ \subseteq B$. Hence

$$A_{B^+}^T y = A_{B^+}^T(y(\pi) + \epsilon\eta(\pi)) = c_{B^+} + \epsilon(e_j)_{B^+}.$$

Also,

$$A_{N^+}^T y = A_{N^+}^T(y(\pi) + \epsilon\eta(\pi)) \leq c_{N^+} + \epsilon A_{N^+}^T\eta(\pi) < c_{N^+} + \epsilon(e_j)_{N^+}.$$

So y is an interior point of $\mathcal{D}(\pi_p^+)$. Hence $(x^+(\pi), y)$ is a strictly complementary pair. So π_p^+ is the optimal partition in the new situation, and as a
consequence, the value of $z(\beta, \gamma + \epsilon)$ is given by (17). This proves the first
part of the theorem. The proof of the second part goes in the same way,
and is therefore omitted.   □

As in the case where β was varied, we have a condition under which π does
not change for small variations in γ.

Theorem 4.7 *If $(e_j)_B$ belongs to the row space of A_B then $x^-(\pi) =
x^+(\pi) = x(\pi)$ and $\pi_p^- = \pi_p^+ = \pi$.*

Proof: If $(e_j)_B$ belongs to the row space of A_B then there exists a vector
v such that $A_B^T v = (e_j)_B$. Now let $x \in \mathcal{P}(\pi)$. Then we may write

$$x_j = (e_j)_B^T x_B = v^T A_B x_B = v^T b,$$

proving that x_j is constant on $\mathcal{P}(\pi)$. But then the central solution of the
problems (13) and (14) is simply the analytic center $x(\pi)$ of $\mathcal{P}(\pi)$. So we
have $x^-(\pi) = x^+(\pi) = x(\pi)$ and $\pi_p^- = \pi_p^+ = \pi$. Thus the theorem follows.
                                                                                   □

A direct consequence of the last two results is

Corollary 4.3 *If $(e_j)_B$ belongs to the row space of A_B then x_j is constant
on $\mathcal{P}(\pi)$. Furthermore, the interval $G_\beta(\pi) := \{\gamma : \pi(\beta, \gamma) = \pi\}$ is open,
and $z(\beta, \gamma)$ is linear on this interval, with slope $x_j(\pi)$.*   □

Theorem 4.8 *If $(e_j)_B$ does not belong to the row space of A_B then $\pi_p^+ \neq \pi$
and $\pi_p^- \neq \pi$. Furthermore, $(e_j)_{B+}$ belongs to the row space of A_{B+} and
$(e_j)_{B-}$ belongs to the row space of A_{B-}. Moreover $x^-(\pi) > x(\pi) > x^+(\pi)$.*

Proof: If $(e_j)_B$ does not belong to the row space of A_B then there exists
no η such that $A_B^T \eta = (e_j)_B$. So for any feasible solution η of the problem
(15) at least one of the inequalities of $A_B^T \eta \leq (e_j)_B$ is strict. From this it
is easy to see that the optimal partition for the problem (15), and hence
also for the problem (13), must differ from π. Clearly, $x(\pi)$ is feasible for
the problem (13). But since π differs from the optimal partition π_p^+, $x(\pi)$
cannot be optimal. Hence we must have $x^+(\pi) < x(\pi)$. Finally, if η^+ is
an optimal solution of (13) then $A_{B+} \eta^+ = (e_j)_{B+}$. Hence $(e_j)_{B+}$ belongs
to the row space of A_{B+}. Just in the same way one shows that π_p^- differs
from π, $x(\pi) < x^-(\pi)$, and that $(e_j)_{B-}$ belongs to the row space of A_{B-}.
Hence the proof is complete.   □

Corollary 4.4 *If* $(e_j)_B$ *does not belong to the row space of* A_B *then the interval* $G_\beta(\pi) := \{\gamma : \pi(\beta, \gamma) = \pi\}$ *is a singleton. Moreover,* $z(\beta, \gamma)$ *has a breakpoint in* γ, *with left derivative* $x^-(\pi)$ *and right derivative* $x^+(\pi)$. $\square$

Summarizing the above results we may state

Theorem 4.9 *If* β *is kept fixed, then* $z(\beta, \gamma)$ *is a piecewise linear concave and monotonic function of* γ. *The breakpoints occur where the optimal partition* $\pi(\beta, \gamma)$ *changes.* $\square$

4.3 Two–sided–ranging

Finally, we will consider the case where both β and γ are allowed to vary. As a matter of fact, the results immediately follow from the analysis above. It can be seen that changes in β have an effect on $\mathcal{P}(\pi)$ and not on $\mathcal{D}(\pi)$. Analogously, changes in γ effect $\mathcal{D}(\pi)$ and not $\mathcal{P}(\pi)$. Hence, the ranges determined for β and γ separately also apply for varying both and we conclude the following.

Theorem 4.10 *Let* (β, γ) *be a feasible pair with optimal partition* π. *Then one of the following holds.*

- *If* e_i *in the column space of* A_B *and* $(e_j)_B$ *in the row space of* A_B *then* $G(\pi)$ *is an open rectangular set.*

- *If* e_i *not in the column space of* A_B *and* $(e_j)_B$ *in the row space of* A_B *then* $G(\pi)$ *is an open interval for* γ *and* β *is constant.*

- *If* e_i *in the column space of* A_B *and* $(e_j)_B$ *not in the row space of* A_B *then* $G(\pi)$ *is an open interval for* β *and* γ *is constant.*

- *If* e_i *not in the column space of* A_B *and* $(e_j)_B$ *not in the row space of* A_B *then* $G(\pi)$ *consists only of the pair* (β, γ). $\square$

We conclude by presenting an example.

Example 4.2 Consider the following problem:

$$\max\{y_1 + \beta y_2 : 0 \le y_1, y_2 \le 2, \ y_1 + y_2 \le \gamma, \ y_1 + y_2 \le 3\}.$$

The feasible region is drawn in Figure 2. Note that this problem is feasible if $\gamma \ge 0$, and bounded for all values of β. If $\gamma = 0$ the interior region is empty, so the pair (β, γ) is feasible only if $\gamma > 0$. Figure 3 shows the optimal partitions for all feasible pairs.

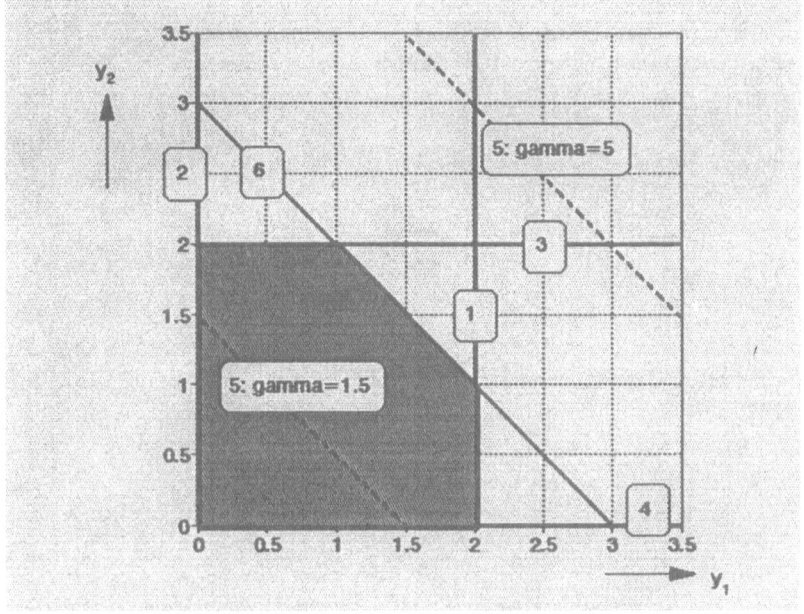

Figure 2: Feasible region for the example.

Acknowledgement

This paper is a result of cooperation with O. Güler (TU Delft), D. den Hertog (TU Delft), B. Jansen (TU Delft), T. Terlaky (TU Delft) and J.-Ph. Vial (University of Genève). For a more extensive treatment of the material in the Sections 2,3 and 4 of this paper the reader may be referred to respectively [12], [24] and [13]. We kindly acknowledge financial support by Shell Research.

References

[1] Anstreicher, K.M (1985), A Monotonic Projective Algorithm for Fractional Linear Programming, *Algorithmica 1*, 483–498.

[2] Anstreicher, K.M., Bosch, R.A. (1988), Long Steps in a $O(n^3L)$ Algorithm for Linear Programming, Preprint, Yale School of Organization and Management, New Haven, CT.

[3] Barnes, E.R. (1986), A Variation on Karmarkar's Algorithm for Solving Linear Programming Problems, *Mathematical Programming* 36, 174–182.

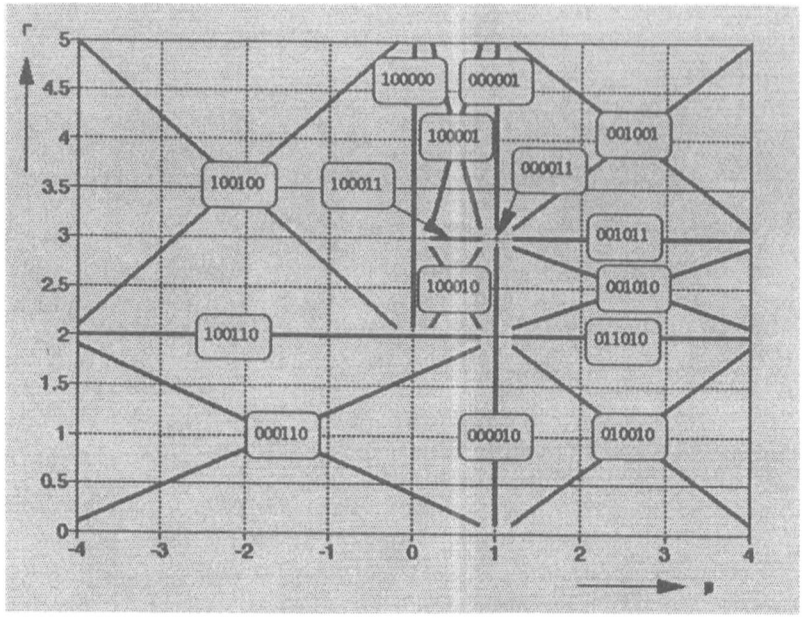

Figure 3: Optimality sets for the example.

[4] Dikin, I.I. (1967), Iterative Solution of Problems of Linear and Quadratic Programming, *Doklady Akademiia Nauk SSSR* 174, 747–748.

[5] De Ghellinck, G. and Vial, J.-Ph. (1986), A Polynomial Newton Method for Linear Programming, *Algorithmica* 1, 425–453.

[6] Freund, R.M. (1988), Polynomial–Time Algorithms for Linear Programming Based only on Primal Scaling and Projected Gradients of a Potential Function, Working Paper 182–88, Massachusetts Institute of Technology, Massachusetts.

[7] Gill, P.E., Murray, W., Saunders, M.A., Tomlin, J.A. and Wright, M.H (1986), On Projected Newton Barrier Methods for Linear Programming and an Equivalence to Karmarkar's Projective Method, *Mathematical Programming* 36, 183–209.

[8] Goldfarb, D. and Liu, S. (1988), An $O(n^3 L)$ Primal Interior Point Algorithm for Convex Quadratic Programming, Manuscript, Department of Industrial Engineering and Operations Research, Columbia University, New York, NY.

[9] Goldman, A. J. and Tucker, A. W. (1956). Theory of linear programming. *Linear Inequalities and Related Systems* (H. W. Kuhn and A. W. Tucker,

eds.), Annals of Mathematical Studies, No. 38, Princeton University Press, Princeton, New Jersey, 53–97.

[10] Gonzaga, C.C. (1989), An Algorithm for Solving Linear Programming Problems in $O(n^3 L)$ Operations, In *Progress in Mathematical Programming, Interior Point and Related Methods*, 1–28, N. Megiddo ed., Springer Verlag, New York.

[11] Gonzaga, C.C. (1990), Polynomial Affine Algorithms for Linear Programming, *Mathematical Programming* 49, 7–21.

[12] Güler, O., Roos, C., Terlaky, T., Vial, J.-Ph. (1992), Interior Point Approach to the Theory of Linear Programming, Cahiers de Recherche No. 1992.3, Faculté des Sciences Economiques et Sociales, Université de Genève.

[13] Jansen, B., Roos, C., Terlaky, T. (1992), Interior Point Approach to Postoptimal and Parametric Analysis in Linear Programming, Working paper.

[14] Karmarkar, N. (1984). A new polynomial–time algorithm for linear programming. *Combinatorica* 4, 373–395.

[15] Kojima, M., Mizuno, S., and Yoshise, A. (1989). A primal–dual interior point algorithm for linear programming. *Progress in Mathematical programming, Interior Point and Related Methods* (N. Megiddo, ed.), Springer Verlag, New York, 29–48.

[16] Kojima, M., Mizuno, S., Yoshise, A. (1989), A Polynomial Time Algorithm for a Class of Linear Complementarity Problems, *Mathematical Programming* 44, 1–26.

[17] McLinden, L. (1980). An analogue of Moreau's proximation theorem, with applications to the nonlinear complementarity problem. *Pacific Journal of Mathematics* 88, 101–161.

[18] McShane, K. A., Monma, C. L., and Shanno, D. F. (1989). An implementation of a primal–dual interior point method for linear programming. *ORSA Journal on Computing* 1, 70–83.

[19] Mizuno, S., Todd, M.J. (1989), An $O(n^3 L)$ Long Step Path Following Algorithm for a Linear Complementarity Problem, Technical Report, School of Operations Research and Industrial Engineering, Cornell University, Ithaca, NY.

[20] Monteiro, R.D.C. and Adler, I. (1989). Interior path following primal–dual algorithms. Part I: Linear programming. *Mathematical Programming* 44, 27–41.

[21] Renegar, J. (1988), A Polynomial–Time Algorithm, Based on Newton's Method, for Linear Programming, *Mathematical Programming* 40, 59–93.

[22] Rockafellar, R. T. (1970). *Convex Analysis.* Princeton University Press, Princeton, New Jersey.

[23] Roos, C. (1987), A New, Trajectory Following Polynomial–Time Algorithm for the Linear Programming Problem, *Journal on Optimization Theory and its Applications* 63, 433–458.

[24] Roos, C. and Vial, J.–Ph. (1988), A Polynomial Method of Approximate Centers for Linear Programming, *Mathematical Programming* 54, No. 3, forthcoming.

[25] C. Roos and J.Ph. Vial (1990). *Long Steps with the Logarithmic Penalty Barrier Function in Linear Programming.* In *Economic Decision–Making: Games, Economics and Optimization* (dedicated to Jacques H. Drèze), edited by J. Gabszevwicz, J.–F. Richard and L. Wolsey, Elsevier Science Publisher B.V., 433–441.

[26] Schrijver, A. (1986). *Theory of Linear and Integer Programming.* John Wiley & Sons, New York.

[27] Sonnevend, Gy. (1985) An "analytic center" for polyhedrons and new classes of global algorithms for linear (smooth convex) programming. In A. Prékopa, J. Szelezsán and B. Strazicky, eds. *System Modelling and Optimization: Proceedings of the 12th IFIP–Conference, Budapest, Hungary, September 1985,* Vol. 84. of *Lecture Notes in Control and Information Sciences,* Springer Verlag, Berlin, 866–876.

[28] Strang, G. (1988). *Linear Algebra and its Applications* (3rd edition). Harcourt Brace Jovanovich Publishers, San Diego, California.

[29] Todd, M.J. and Burrell, B.P. (1986), An Extension of Karmarkar's Algorithm for Linear Programming Using Dual Variables, *Algorithmica* 1, 409–424.

[30] Todd, M. J. and Ye, Y. (1990). A centered projective algorithm for linear programming. *Mathematics of Operations Research* 15, 508–529.

[31] Vaidya, P.M. (1990), An Algorithm for Linear Programming which Requires $O(((m + n)n^2 + (m + n)^{1.5}n)L)$ Arithmetic Operations, *Mathematical Programming* 47, 175–201.

[32] Vanderbei, R.J., Meketon, M.S., Freedman, B.A. (1986), A Modification of Karmarkar's Linear Programming Algorithm, *Algorithmica* 1, 395–407.

[33] Ye, Y. (1987), Interior Algorithms for Linear, Quadratic and Linearly Constrained Convex Programming, Ph.D. Dissertation, Engeneering–Economic Systems Department, Stanford University, Stanford.

Some reflections on Newton's method

F. Twilt
Faculty of Applied Mathematics
University of Twente
P.O. Box 217, 7500 AE Enschede, The Netherlands

Abstract

An infinitesimal version (Newton flow) of Newton's iteration for finding zeros of rational functions on the complex plane is introduced. Structural stability aspects are discussed, including a characterization and classification of structurally stable Newton flows in terms of certain plane graphs. Possible generalizations to other classes of functions are indicated and several open problems are posed.

Keywords: Newton flow, structural stability, rational function.
1991 Mathematical subject classification: 05C10, 30C15, 34D30, 49M15.

1. Introduction and Motivation

Newton's method and those methods based on it are of great practical use when dealing with the problem of solving nonlinear equations.

We begin with a brief explanation of this method. Let f be a mapping from the Euclidean space $\mathbf{R}^k$ to itself and let ω be a zero for f. Suppose that f is continuously differentiable and that the Jacobi matrix of f (denoted by $Df(\cdot)$) is nonsingular at ω. Now, let $x_0 \in \mathbf{R}^k$ be given. Then, the so-called *Newton iteration scheme*

$$x_{n+1} = x_n - (Df(x_n))^{-1} f(x_n), \qquad n = 0, 1, \ldots, \tag{1}$$

gives rize to a well-defined sequence (x_n), which converges to ω, provided that the starting point x_0 is chosen sufficiently close to ω. For a proof, including an estimation of the (very satisfactory) rate of convergence, we refer to [I-1]. Here we merely note that, under the above conditions, x_{n+1} is the zero for the linear approximation of f around x_n, rather than a zero for f itself; see Fig. 1 for a

A. van der Burgh and J. Simonis (eds.), Topics in Engineering Mathematics, 217–237.
© 1992 *Kluwer Academic Publishers.*

geometrical interpretation in the one dimensional case.

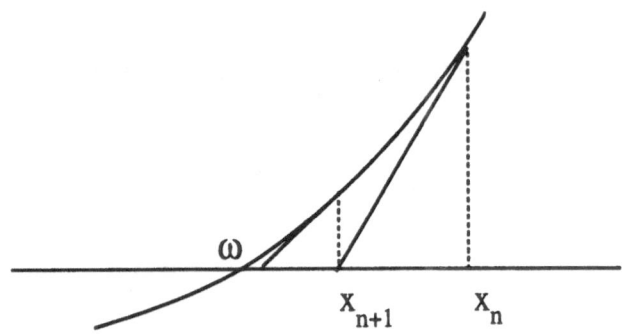

Figure 1.

Newton's method is frequently used in many practical situations. For instance, in Optimization (cf. [I-2]) where f is either the gradient of the object function (unconstrained optimization) or f arizes from the Kuhn-Tucker optimality conditions (constrained optimization). Another typical application is the numerical solution of the corrector equation, when dealing with stiff differential equations, cf. [I-3], [A-1].

Even in the simple situation as depicted in Fig. 1, it is evident that the Newton iteration (1) is easy to understand only *locally* around a zero such as ω, whereas its *global behavior* might be very wild and unsurveyable. In order to enlarge the neighborhood where the starting point x_0 should be chosen, it is usual to control each iteration step by aid of a positive damping factor h_n:

$$x_{n+1} = x_n - h_n \cdot (Df(x_n))^{-1} f(x_n), \qquad n = 0, 1, \ldots . \tag{2}$$

Choosing the damping factors h_n in (2) small and taking the difference quotient $(x_{n+1} - x_n)/h_n$ into account, we arrive at the so-called *continuous Newton method*:

$$\frac{dx}{dt} = -(Df(x))^{-1} f(x), \qquad x(0) = x_0, \tag{3}$$

which may be seen as an infinitesimal version of (1).

Conversely, the Euler discretization with steplength h_n, applied to the differential equation (3) yields the damped Newton iteration (2); compare also [I-4]. Hence, if we are interested in the *global behavior* of the damped Newton method (2), it is reasonable to study the global features of the phase portrait of (3) (i.e. the family of its maximal trajectories). Especially, when h_n is small and system

(3) behaves *structural stable* (see below), such study will be relevant. This is the basic motivation behind the work on the continuous Newton method as done by several authors, e.g. Braess [A-2], Branin [A-3], Diener [A-4, A-5], Garcia and Gould [A-6], Hirsch and Smale [A-7], Jongen, Jonker and Twilt [A-8, A-9, A-10], [I-5], Keller [A-11], Shub, Tischler and Williams [A-12], Smale [A-13, A-14].

The study of the continuous Newton method is the aim of this present paper. In order not to blow up the size of the presentation we will focus on a special class of functions f, namely rational functions on a complex variable. Doing so, the equation (3) becomes a dynamical system on the (complex) *plane* C and moreover, the whole machinery of complex function theory can be used. Therefore, in this case the theory will be relatively simple and the results take a very sophisticated form. However, we emphasize that a large part of the results obtained in this special case remains valid for more general classes of functions f, be it often in a weaker form (see [I-5], Ch.9) and also several remarks throughout the text). In the last section we pose some open questions, especially with respect to more general classes of functions.

This paper accentuates the basic ideas and the (geometrical) evidence of the results, rather than to dwell on technical details. These details can be found in the references which are subdivided into an introductory ([I]) and an advanced ([A]) level.

2.   Rational Newton Flows

Let f be a non-constant *rational* function of a complex variable z (thus, $f = p/q$ with p and q relatively prime polynomials). Then, system (3) takes the form

$$\frac{dz}{dt} = -\frac{f(z)}{f'(z)}, \qquad z(0) = z_0, \tag{4}$$

where f' stands for the derivative with respect to z. This autonomous differential equation (and frequently also the vector field given by the right hand side of it) is called *rational Newton flow*, and will be denoted by $\mathcal{N}(f)$.

Let $N(f), P(f)$ be the set of *zeros*, respectively *poles* for f. So, $N(f) = \{z \mid p(z) = 0\}$ and $P(f) = \{z \mid q(z) = 0\}$. Moreover, we define $C(f) = \{z \mid f'(z) = 0, f(z) \neq 0\}$ as the set of *critical points* for f. Since f is a rational function, the mutually disjoint sets $C(f)$, $N(f)$ and $P(f)$ are finite. The union $C(f) \cup N(f) \cup P(f)$ contains all possible singularities for $\mathcal{N}(f)$ (i.e. points where $\mathcal{N}(f)$ fails to be complex analytic). Let z_0 be such a singularity. Then, if $z_0 \in (N(f) \cup P(f))$, this singularity can be removed by defining $f(z_0)/f'(z_0) = 0$, whereas, if $z_0 \in C(f)$ the singularity is *not* removable. Therefore, we may regard $\mathcal{N}(f)$ as a (*complex*) *analytic vector field* on $C \backslash C(f)$ with finitely many equilibria (zeros), namely the points of $N(f) \cup P(f)$.

Consequently, the maximal trajectory [1] of $\mathcal{N}(f)$ through a point $z_0 \notin C(f)$ is well-defined, either as a regular curve (if $z_0 \notin N(f) \cup P(f)$) or as the singleton $\{z_0\}$ (if $z_0 \in N(f) \cup P(f)$).

The following basic observation is easily verified:

$$\mathcal{N}(f) = -\mathcal{N}(1/f). \tag{5}$$

Moreover, for $z_0 \notin C(f)$ direct integration of (4) yields

$$f(z(t)) = f(z_0)e^{-t}, \tag{6}$$

where $z(t)$ represents a solution of (4). Consequently, $|f(z(t))|$ depends strictly decreasing on t and moreover, on the trajectories of $\mathcal{N}(f)$ the argument of $f(z)$ remains *constant*. So, a trajectory of $\mathcal{N}(f)$ through $z_0 \notin (N(f) \cup P(f))$ is contained in the inverse image under f of the half ray $\arg w = \arg f(z_0)$. So, from the elementary properties of (multifold) conformal mappings (cf. [I-6]) one easily concludes that the local phase portrait around a point $\check{z}$ of $C(f) \cup N(f) \cup P(f)$ is of one of the types as depicted in Fig.2. The last two pictures in Fig. 2 treat critical points which are 1-fold respectively 2-fold zeros for f'.

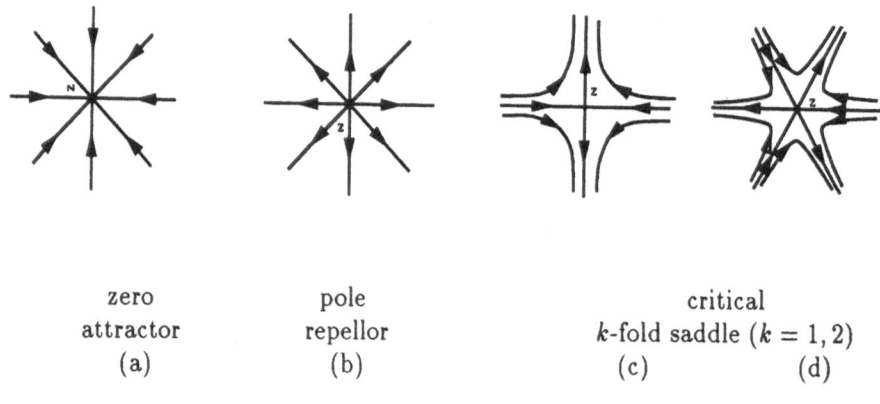

zero	pole	critical
attractor	repellor	k-fold saddle ($k = 1, 2$)
(a)	(b)	(c) (d)

Figure 2

In Fig. 2-a (Fig. 2-b) for *every* $\Theta \in (-\pi, \pi]$, a *unique* trajectory γ exists such that if z tends to $\check{z}$ along γ, then $\arg(z - \check{z})$ tends to Θ. In Fig. 2-c,d the angle between two subsequent trajectories, one of which tending to $\check{z}$, the other leaving from $\check{z}$, equals $\pi/(k+1)$. In the case of a simple (= 1-fold) critical point $\check{z}$, see Fig. 2.c the *two* maximal trajectories "leaving" from $\check{z}$, together

---

[1] A maximal trajectory of $\mathcal{N}(f)$ through z_0 is an integral curve of $\mathcal{N}(f)$ of the type

$$z(t), \ t \in J \ (= \text{interval around } t = 0), \ z(0) = z_0,$$

with the property that any other integral curve of this type is defined on a subinterval of J

with $\check{z}$ constitute the so-called *unstable manifold*; the two maximal trajectories "tending" to $\check{z}$ define the *stable manifold* at $\check{z}$.

Note that Fig. 2-a is in accordance with the convergence behavior of the affliated discrete Newton iteration and that Fig. 2-b illustrates relation (5). We emphasize that $\mathcal{N}(f)$ is *not* defined at the critical points for f.

Remark 1.

1.1 The above classification of local phase portraits remains true if f is a (non-constant) meromorphic function (i.e. a complex function which attains as possible singularities *only poles*).

1.2. In the situation of Section 1 (f is a continuously differentiable mapping from $\mathbf{R}^k$ to $\mathbf{R}^k$ and ω is a zero for f) one easily proves: the local phase portrait of system (3) around ω has the same structure as depicted in Fig. 2.a. as soon as system (3) is well-defined at ω (i.e. the Jacobian matrix $(Df(\omega))$ is non-singular). Note that the latter condition reduces to $f'(\omega) \neq 0$ when f is a rational function on $\mathbf{C}$.

When turning over to global aspects, we encounter the problem that $\mathcal{N}(f)$ is singular on $C(f)$. We overcome this problem by considering the following vector field on $\mathbf{C}$ (where $\bar{f}'(z)$ stands for complex conjugate):

$$\bar{\mathcal{N}}(f)_{|z} = \begin{cases} -(1 + |f(z)|^4)^{-1} \cdot \bar{f}'(z) \cdot f(z), & \text{if } z \notin P(f); \\ 0, & \text{if } z \in P(f). \end{cases} \tag{7}$$

Apparently, on $\mathbf{C}\backslash P(f)$, the vector field $\bar{\mathcal{N}}(f)_{|z}$ depends *real* analytically on the variables $x(= Re\ z)$ and $y(= Im\ z)$ but *not* complex analytically on z. Moreover, its equilibria are the points in $C(f) \cup N(f) \cup P(f)$. Since, outside the latter set, $\mathcal{N}(f)$ and $\bar{\mathcal{N}}(f)$ are equal, up to the strictly positive factor $(1 + |f(z)|^4)^{-1} \cdot |f'(z)|^2$, the phase portraits of these two flows are the same on $\mathbf{C}\backslash C(f)$, including the orientations of the trajectories (cf. Fig. 3). By inspection, one easily verifies that

$$\bar{\mathcal{N}}(f) = -\bar{\mathcal{N}}(1/f). \tag{8}$$

From this relation it follows that $\bar{\mathcal{N}}(f)$ is also real analytic in $P(f)$.

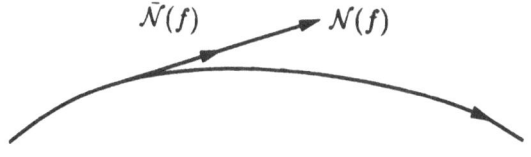

Figure 3

Remark 2.

2.1. Note that after the above "desingularization" in the underlying discrete Newton method only the damping factors have been changed.

2.2. The "desingularization" procedure also works for general meromorphic functions.

2.3. A similar desingularization (occasionally reversing orientations) is possible for *all* systems (3). In fact, this is the usual starting point for work on continuous Newton methods.

As it is well-known, a rational function can be extended, using the transformation $w = 1/z$, to the Riemannian sphere S^2. So, it is not surprising that rational Newton flows induce vector fields on S^2. From a geometrical point of view, the relation between $\bar{\mathcal{N}}(f)$ and the induced "spherical" Newton flow – say $\bar{\bar{\mathcal{N}}}(f)$ – is very simple: the maximal trajectories of $\bar{\mathcal{N}}(f)$ are mapped onto those of $\bar{\bar{\mathcal{N}}}(f)$ basically by means of a *stereographic projection*. For the actual construction of $\bar{\bar{\mathcal{N}}}(f)$ we refer to [A-8]. Here, we merely give the technical formulation of the "extension" result:

Given $\bar{\mathcal{N}}(f)$, then a (real analytic) vector field – say $\bar{\bar{\mathcal{N}}}(f)$ – on S^2 exists, such that with respect to the z-chart (to be identified with C) the phase portraits of $\bar{\mathcal{N}}(f)$ and $\bar{\bar{\mathcal{N}}}(f)$ are the same. Moreover, at $z = \infty$ ("north pole") $\bar{\bar{\mathcal{N}}}(f)$ exhibits an atractor (respectively, a repellor, a k-fold saddle) when at $w = 0$ the function $f(1/w)$ attains a zero (respectively, a pole, a k-fold critical point).

The above result is very usefull. In fact, it enables us to apply elements of the theory on (real analytic) vector fields with a 2-dim. *compact* support [I-7] to rational Newton flows. For example, the following characterization of the limiting sets of $\mathcal{N}(f)$-trajectories is a direct consequence of Bendixon's theorem on S^2 (cf. [I-7]) applied to $\bar{\bar{\mathcal{N}}}(f)$:

Let the maximal $\mathcal{N}(f)$-trajectory through a point z_0 be given by $z(t)$. Then, we have: $z(t)$ ends up (when t increases/decreases) in a point of $C(f) \cup N(f) \cup P(f)$, or possibly in $z = \infty$.

Remark 3. The behavior of limiting sets of trajectories of system (3) is a main topic in the general theory on the continous Newton method. Unfortunately, a transfer to vector fields with compact support is only possible in certain very special situations (such as the case where f is a rational function on C) Nevertheless, there are some (partial) results (see [I-5] and [A-13]).

3. Structural Stability

As in section 2, let f be a non-constant rational function. We represent f as $f = p_n/q_m$, where p_n and q_m are two polynomials (of degree n and m respectively) which are relatively prime.

For the sake of simplicity, we put $n > m$. Note however, that, in view of (5) and (8), the case $n > m$ already covers the case $n < m$. So, our simplified presentation essentially only disregards the case $n = m$.

The set of all rational functions f, with $n > m$ is denoted by R_+.

Roughly speaking, a rational Newton flow is called *structurally stable* if the qualitative behavior of the phase portrait of $\mathcal{N}(f)$ does not change under small perturbations of the coefficients of f. In order to make this precise, we need the concept of (topological) equivalence between flows on $\mathbf{C}$:

Let $f_1, f_2 \in R_+$. Then, the flows $\mathcal{N}(f_1)$ and $\mathcal{N}(f_2)$ are called *equivalent* ($\sim$) if a homeomorphism from $\mathbf{C}$ onto itself exists, mapping the maximal trajectories of one flow onto those of the other. (Informally this means that both flows have the same behavior).

The flow $\mathcal{N}(f)$, $f \in R_+$, is called *structurally stable* if after sufficiently small, but for the rest arbitrary, perturbations of the coefficients of f the resulting functions give rize to Newton flows which are equivalent with $\mathcal{N}(f)$.

Now we ask for a characterization of those functions $f \in R_+$ for which $\mathcal{N}(f)$ is structurally stable. To this end we introduce the concept of a non-degenerate function:

The function $f \in R_+$ is called *non-degenerate* if:

1. all zeros, poles and critical points for f are simple;

2. no two critical points are "connected" by a $\mathcal{N}(f)$- trajectory.

The set of all non-degenerate critical points in R_+ is denoted by $\tilde{R}_+$.

The first condition means: the polynomials p_n, q_m and $p'_n q_m - p_n q'_m$ have only simple zeros. One might argue that the non-degeneracy condition 2 is a condition on $\mathcal{N}(f)$ rather than on f. Note however, that – in view of (6) – for two critical points z_1, z_2 to be *not* connected by a $\mathcal{N}(f)$-trajectory, it is sufficient that $\arg f(z_1) \neq \arg f(z_2)$.

Suppose that f is *degenerate*, i.e. at least one of the above non-degeneracy conditions is violated. Then, even the slightest perturbation on the coefficients of f may cause a dramatic change in the phase portraits of the corresponding flows (e.g. because a double zero splits into two simple zeros, or because saddle connections are broken up). Thus, we may expect a close relationship between the concepts of (non-)degeneracy for f and structural stability for $\mathcal{N}(f)$. This relationship is the content of the main theorem of this section:

Theorem 1. Let f be a function in R_+. Then $\mathcal{N}(f)$ is structurally stable if and only if f is non-degenerate.

Sketch of the proof.

By the aid of a classical result due to de Baggis and Peixoto (cf. [I-8], [A-15]), applied to the *spherical* Newton flow $\bar{\bar{\mathcal{N}}}(f)$, it is possible to prove ([A-8]) that $\mathcal{N}(f)$ is structurally stable if and only if the following three conditions hold:

(i) $\mathcal{N}(f)$ does not exhibit periodic (i.e. closed) trajectories;

(ii) All equilibria of $\bar{\mathcal{N}}(f)$ are *hyperbolic*[2];

(iii) No two critical points for f are "connected" by a $\mathcal{N}(f)$-trajectory.

In view of (6), the condition (i) holds for *all* rational Newton flows. By inspection of (7), it is easily verified that the condition (ii) is equivalent with the non-degeneracy condition 1. Finally, condition (iii) is nothing else than the non-degeneracy condition 2.   □

We conclude this section by discussing a result, which – roughly speaking – states that non-structurally stable rational Newton flows are the *very exceptions*.

It will be intuitively plausible that, if for some $f \in R_+$ the non-degeneracy conditions hold, then this remains true for any function which results from f by sufficiently small perturbations of its coefficients. So, non-degeneracy is a *stable* property. On the other hand, any $f \in R_+$ can be *approximated* (using arbitrarily small perturbations of its coefficients) by non-degenerate functions in R_+. These observations are properly formulated in the following "genericity" lemma:

Lemma 1. The set $\widetilde{R}_+$ is τ-open and τ-dense in R_+.

Here, τ is a *topology* on R_+, which is natural in the following sense: Let $f = p_n/q_m$ be a function in R_+. Given $\epsilon > 0$ sufficiently small, then there exists a τ-open neighborhood Ω of f, such that any $g \in \Omega$ can be represented by $\widetilde{p}_n/\widetilde{q}_m$ such that the coefficients of $\widetilde{p}_n$ and $\widetilde{q}_m$ are in ϵ-neighborhoods of the corresponding coefficients of p_n and q_m. For a precise definition of τ and for a proof of the above lemma, we refer to [A-8].

Remark 4.

4.1. In order to get rid of the restriction $n > m$, we have to define the concepts of "structural stability" and "non-degeneracy" in terms of the *spherical* Newton-flows $\bar{\bar{\mathcal{N}}}(f)$.

4.2. The concepts of "structural stability" and "non- degeneracy" can also be defined for the general Newton systems (3). Under a so-called "global boundary" condition, characterization and genericity results analoguous to Theorem 1 and Lemma 1, holds in the *2-dimensional* case (cf. [I-5]).

---

[2]An equilibrium of $\mathcal{N}(f)$ (regarded as a vector field on $\mathbf{R}^2$) is called hyperbolic, if the Jacobi matrix of $\mathcal{N}(f)$ at this equilibrium has only eigenvalues with non vanishing real parts.

4. Newton Graphs

Let f be a rational function in $\tilde{R}_+$ (see Section 3). So, $\mathcal{N}(f)$ is structurally stable. We shall characterize the qualitative behavior of the global phase portrait of $\mathcal{N}(f)$ in terms of a certain *plane graph* (i.e. the representation of an abstract graph drawn in the plane). This plane graph will be called $G(f)$, and is defined as follows:

- the vertices of $G(f)$ are the zeros for f;

- the edges of $G(f)$ are the unstable manifolds (cf. Section 2) at the critical points for f.

Note that $G(f)$ is a "part" of the phase portrait of $\mathcal{N}(f)$. As an example, see Fig. 4, where f is of the form p_4/q_1, and the zeros and pole for f are denoted by $\omega_1, \ldots, \omega_4$ and by α respectively; two critical points are indicated by σ_1, σ_2.

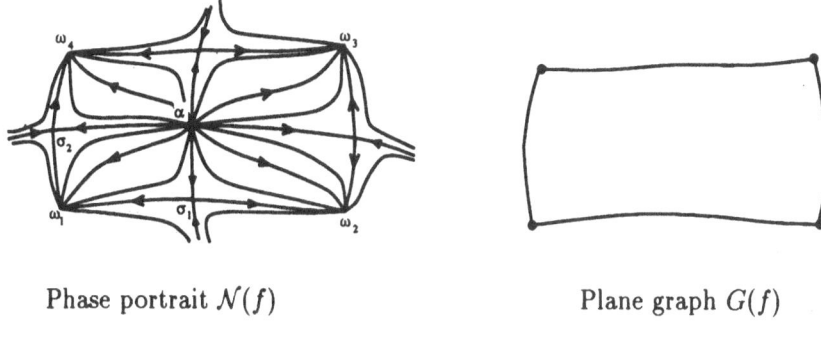

Phase portrait $\mathcal{N}(f)$ Plane graph $G(f)$

Figure 4

When all edges of a given plane graph G are deleted from the plane, the connected components of the resulting set are called the *faces* of G. For a plane graph $G(f)$, it can be proved ([A-9]) that each bounded face contains precisely one pole for f. In fact, such a face is completely "filled up" with $\mathcal{N}(f)$-trajectories leaving from this pole (cf. Fig. 4).

The plane graph $G(f)$ has the following properties:

(i) At any vertex of $G(f)$ the angle between two edges with this vertex in common is well-defined and never vanishes.

(ii) For any bounded face of $G(f)$, the sum of all angles spanning a sector of this face at the vertices, is equal to 2π. (In this paper, an angle is always counted positively).

(iii) $G(f)$ is connected.

We indicate the proof of this assertion in the situation of Fig. 4:

The edges of $G(f)$ are built up by $\mathcal{N}(f)$-trajectories. So, property (i) follows directly from the analysis of the local phase portrait of $\mathcal{N}(f)$ around a zero for f (see Section 2, especially the comment to Fig. 2.a).

In Fig. 4, the unstable manifolds through σ_1 and σ_2 constitute an angle at ω_1, whereas the stable manifolds through σ_1 and σ_2 form an angle at α. These angles are equal. This is easily seen by observing that $\arg f(z)$ remains constant on $\mathcal{N}(f)$-trajectories [cf. (6)] and that, f being non-degenerate, the functions f, respectively $1/f$ are conformal (thus *angle preserving*) at ω_1, respectively α. In a similar way each angle spanning a sector (at a vertex) of the $G(f)$-face determined by α is found back as one of the the angles at α between two stable manifolds through critical points on the boundary of this face. Since the sum of all such angles at α apparently equals 2π, we have proved – in the situation of Fig. 4 – property (ii). In the general case, the proof runs along the same lines, but requires a precise study of the boundaries of the $G(f)$-faces. This will be ommitted here (see [A-9]).

Finally, the connectedness of $G(f)$ is a direct consequence of the Poincaré-Hopf index theorem [I-9] applied to the desingularized vector field $\bar{\mathcal{N}}(f)$, as well as the Euler-polyhedron formula for plane graphs ([I-10]) applied to $G(f)$.   □

As usual, two plane graphs are called (topological) equivalent ($\sim$) if a homeo-morphism of the plane onto itself exists which maps the edges and vertices of one graph onto those of the other.

Any plane graph for which the above properties (i), (ii) and (iii) are fulfilled is called a *Newton graph*. So, $G(f)$ is a Newton graph. Now the crucial point is that, up to topological equivalency, the converse is also true and moreover, the underlying Newton graphs distinguish between equivalency of the corresponding Newton flows. In fact, we have:

Theorem 2.

  a. Let G be a Newton graph. Then, a structurally stable Newton flow $\mathcal{N}(f)$, $f \in \tilde{R}_+$, exists such that $G \sim G(f)$.

  b. Let f_1 and f_2 be two rational functions in $\tilde{R}_+$. Then, $\mathcal{N}(f_1) \sim \mathcal{N}(f_2)$ if and only if $G(f_1) \sim G(f_2)$.

Concerning the proof [A-9] of this *classification* theorem, we merely give the following brief comment:

Assertion a. is proved by induction on the number of G-faces and G-vertices. Given $\tilde{f} \in \tilde{R}_+$, the very heart of the proof is the *construction* of a *non-degenerate* f such that, up to equivalency, $G(f)$ results from $G(\tilde{f})$ by adding to it either only one face (but no vertices) or only one vertex (but no faces), thereby effectuating changes in the $G(\tilde{f})$-angles which may be kept arbitrary small. (For the latter claim, the structural stability of $\mathcal{N}(\tilde{f})$ is needed).

Assertion b. is a direct consequence of Peixoto's classification of 2-dim. structurally stable, continuously differentiable vector fields [A-16], applied to the spherical flows $\bar{\bar{\mathcal{N}}}(f)$, $f \in \tilde{R}_+$. □

Remark 5.

5.1. In the formulation of Theorem 2 we can get rid of the condition $n > m$, by treating the "spherical" view point (compare Remark 4.1).

5.2. For the above classification theorem it is essential that $\bar{\bar{\mathcal{N}}}(f)$ is a *two-dimensional* vector field on a *compact* manifold with *finitely* many equilibria.

This section is concluded by presenting two computer drawn pictures of rational Newton flows, one of which is structurally stable (Fig. 5), the other is not (Fig. 6).

We briefly explain the examples as exhibited in Fig. 5 and 6. Apparently, the function f which is treated in Fig. 5 has three *simple* zeros (for $z = \pm 1$ and $z = 5$) and one pole (for $z = 0$). By elementary means, one shows that f has three *simple* critical points (namely the zeros for $2z^3 - 5z^2 - 5$), one critical point being real, the two others non real and complex conjugate. Since the coefficients of f are real, both f and f' are symmetric with respect to the real axis. Hence, such symmetry property also holds for the phase portrait of $\mathcal{N}(f)$. Using this symmetry and the fact that all zeros for f are simple, a carefull analysis of the limiting sets of $\mathcal{N}(f)$-trajectories (cf. Section 2) yields: the critical points for f are *not* connected by $\mathcal{N}(f)$-trajectories. So, it is possible to define a plane graph with as vertices the zeros for f and as edges the unstable manifolds through the critical points. Conclusions of the same type can be drawn with respect to the example in Fig. 6. However, there is one important difference between the two examples. For $z = 0$, the function f (in Fig. 5) has a *simple* pole, whereas h (in Fig. 6) has for $z = 0$ a *double* pole. So, the function f is non-degenerate, and h is degenerate. If follows that, although both flows determine a plane graph "on the zeros and unstable manifolds" *only* in Fig. 5 this flow is structurally stable and the associated graph a Newton graph.

The above analysis is confirmed by the examples in Fig. 5 and 6, where we obtained the phase portraits by numerical following of the trajectories of the Newton flows. The graphs on "the vertices and unstable manifolds" (the dotted lines in the phase portraits) are drawn separately on the right hand sides of the figures. Note that in Fig. 6, the angle spanning a sector of the bounded face at the vertex in $z = 1/5$ already equals 2π. Hence, the graph in Fig. 6 cannot be a Newton graph.

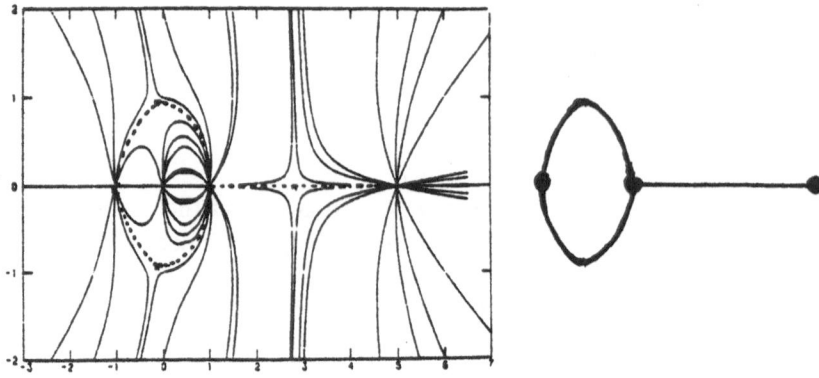

$\mathcal{N}(f)$ $G(f)$ (Newton graph)

$$f(z) = \frac{(z^2-1)(z-5)}{z}$$

Figure 5

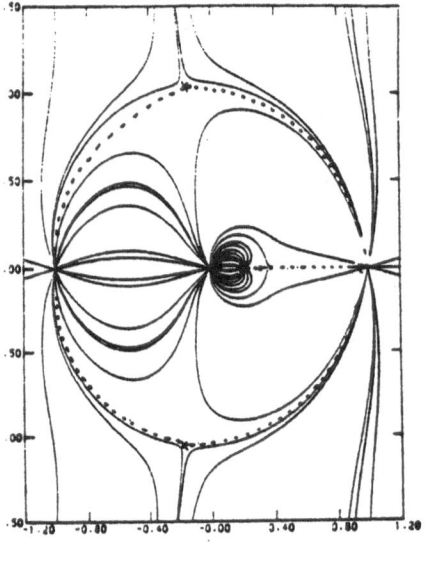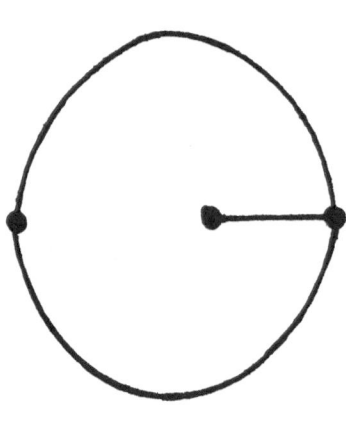

$\mathcal{N}(h)$ $G(h)$ (not a Newton graph)

$$h(z) = \frac{(z^2-1)(5z-1)}{z^2}$$

Figure 6

5. Characterization of Newton Graphs

From a graph theoretical point of view, the definition of Newton graph is not very satisfactory (since it deals with the concept of "angle"). However, one easily derives a necessary condition of purely combinatorial nature, for a connected plane graph G to be equivalent with a Newton graph. To this aim, let C be an arbitrary *cycle* in G, and introduce the integers: $n(C)$ = number of G-vertices inside C but not on C; $\ell(C)$ = number of G-vertices on C; $r(C)$ = number of G-faces inside C. For example, in Fig. 7, we have $n(C) = 1$, $\ell(C) = 3$ and $r(C) = 3$.

The inward, respectively outward, angle at a G-vertex of C is the angle between two consecutive edges of C, spanning a sector of the interior respectively of the exterior of C. The sum of all *inward* angles at the G-vertices on C is denoted by $\sum^i(C)$, whereas $\sum^o(C)$ stands for the sum of all *outward* angles at the G-vertices on C.

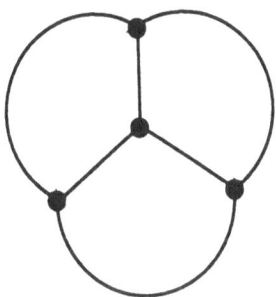   

Plane graph G Cycle C of G

Figure 7

For a *Newton graph* G we have:

$$\sum^i(C) = 2\pi(r(C) - n(C)) \qquad > 0$$
$$\sum^o(C) = 2\pi(\ell(C) + n(C) - r(C)) \quad > 0$$

(This is easily verified by observing that at each vertex the sum of all angles equals 2π and using the Newton graph properties (i), (ii); compare also Fig. 7).

Consequently, a *necessary* condition for G in order to be equivalent with a Newton graph is that the following inequalities do hold:

$$n(C) < r(C) < n(C) + \ell(C), \qquad \text{for all cycles } C \text{ in } G \qquad (*)$$

The latter condition turns out to be sufficient as well:

Theorem 3. A connected plane graph is equivalent with a Newton graph if and only if the inequalities (∗) are fulfilled.

For an illustration of this theorem, see Fig. 5 and Fig. 6. The proof of Theorem 3 can be found in [A-10]. Here, we merely indicate the idea behind the sufficiency part in the situation of Fig. 8, where G apparently fulfils the inequalities (∗).

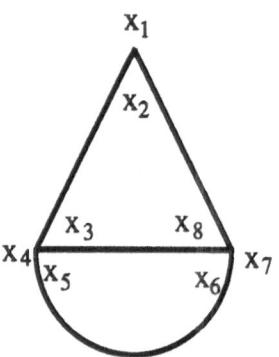

Figure 8

So, let us denote by x_i, $i = 1, \ldots, 8$ the angles of G in Fig. 8. Then, G is equivalent with a Newton graph iff the following system of (in)equalities is fulfilled:

$$
\begin{cases}
x_1 + x_2 & = 2\pi \\
x_3 + x_4 + x_5 & = 2\pi \\
x_6 + x_7 + x_8 = 2\pi \\
x_2 + x_3 \qquad\qquad\quad + x_8 = 2\pi \\
x_5 + x_6 = 2\pi \\
x_i > 0,\ i = 1, \ldots 8
\end{cases}
\tag{9}
$$

Putting $x_i = 2\pi X_i / X_9$, for $i = 1, \ldots, 8$, and $X_9 > 0$, we find for (9):

$$
\begin{cases}
X_1 + X_2 \qquad\qquad\qquad\quad - X_9 = 0 \\
X_3 + X_4 + X_5 \qquad\quad - X_9 = 0 \\
X_6 + X_7 + X_8 - X_9 = 0 \\
X_2 + X_3 \qquad\qquad\quad + X_8 - X_9 = 0 \\
X_5 + X_6 \qquad - X_9 = 0 \\
X_i > 0,\ i = 1, \ldots, 9.
\end{cases}
\tag{10}
$$

The latter system is of the form

$$
\begin{cases}
B \cdot X = 0 \\
\quad X_i > 0,
\end{cases}
\tag{11}
$$

where $X = (X_1, \ldots, X_9)$ and B is the matrix on the coefficients of the equaltities in (10). Due to Stiemke's theorem (cf. [I-11]), system (11) has a solution iff the following system (12) does *not* have a solution for which at *least one* of the inequalities is *strict*:

$$B^T \cdot Y \geq 0, \qquad \text{with } Y = (Y_1, \ldots, Y_5). \tag{12}$$

One easily verifies (since, basically due to (∗), there are more G- vertices than G-faces) that the condition (12) is *only* fulfilled for $Y = 0$. Hence, system (11) and thus also (9) is solvable, i.e. G is a Newton graph. In the general case we proceed in a similar way. Then, the "Stiemke's alternative" for a connected plane graph to be Newtonian is violated by means of the inequalities (∗), in combination with Hall's theorem on distinct representatives [I-12].   □

Now, let n and m be fixed. The above characterization theorem for Newton graphs, together with Theorem 2 yields the possibility to produce complete lists of all – up to equivalency – different structurally stable flows $\mathcal{N}(f)$, with $f = p_n/q_m$.

For example, let f be a polynomial, i.e. $m = 0$. Then, f does not exhibit (finite) poles. Hence, the Newton graph $G(f)$ has no bounded faces (and thus no cycles) i.e. $G(f)$ is a so-called *plane tree*. On the other hand, because of the absence of cycles, each plane tree trivially fulfils the inequalities (∗). It follows that each plane tree represents a structurally stable *polynomial* Newton flow, and vice versa. Moreover, it is possible (cf. [A-17]) to classify – up to equivalency – all plane trees according to their order (= number of vertices). Hence, we also have a complete classification of all structurally stable *polynomial* Newton flows. See for instance Fig. 9, where the case $n = 7$ is treated.

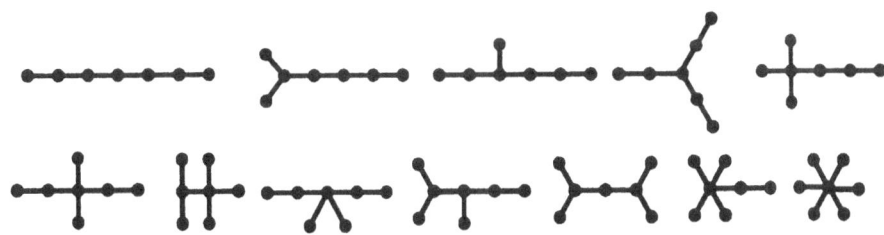

Plane trees representing all structurally stable $\mathcal{N}(p_7)$

Figure 9

Another illustration or our classification result is given in Fig. 10 where the *four* plane graphs are depicted, representing *all* possible structurally stable

Newton flows of the form $\mathcal{N}(f)$, with $f = p_3/q_2$.

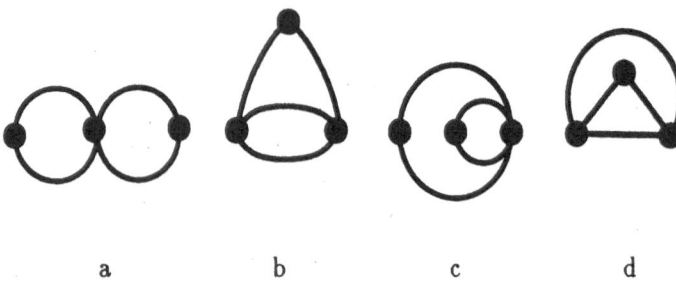

a   b   c   d

Plane graphs representing all structurally stable $\mathcal{N}(p_3/q_2)$

Figure 10

6.   Open Problems

In this final section, we discuss some unsolved problems. Although, occasionally, preliminary results have been obtained, the answers to these questions are still open. As such, the below problems may be the subject of further investigations.

6.1. Elliptic Newton flows

Let f be a non-constant *elliptic* (i.e. a *doubly periodic* and *meromorphic*) function on C. As it is well-known (cf. [I-13]) there is a certain analogy between the theories on rational and on elliptic functions. For example, elliptic functions (with fixed periods) may be seen as functions defined on a *torus* T (compare the Riemannian sphere S^2 in the case of rational functions). Moreover, $\mathcal{N}(f)$ $(= -f(z)/f'(z))$ is also elliptic (with the same periods as f). Hence, the desingularized "elliptic" Newton flow for f may be regarded as a vector field on the *compact* torus T, with only *finitely* may equilibria (cf. Remarks 1.1 and 2.2). So, it is reasonable (cf. Remark 5.2) to ask for a description of the global phase portraits of elliptic Newton flows, especially of those behaving "structurally stable". In the latter case, one may hope for a classification in terms of certain graphs which are drawn on the torus. Up till now, only in simple cases the phase portraits of elliptic Newton flows are understood, e.g. for the Weierstrass $\mathcal{P}$-functions (cf. [A-18]). As an example of a (computer drawn) elliptic Newton flow, see Fig. 11, where the Weierstrass $\mathcal{P}$-function with periods $0.25 + 1.25i$ and 1 is treated.

6.2. A special class of rational Newton flows

As before, by p_n we denote complex polynomials of degree n. I. Diener ([private communication]) proposed a path following method in order to detect by means of only one procedure all zeros of both p_n and its derivative p'_n. Following this idea, we studied rational Newton flows of the special form $\mathcal{N}(p_n/p'_n)$. Here, the question arizes whether it is possible to describe the structurally stable flows of this special type, in terms of Newton graphs. In case $n = 3$, we were able to prove that only plane graphs as depicted in Fig. 10–a,b occur [A-19]. For $n > 3$ the question is still open.

Figure 11

6.3. A special class of non-meromorphic Newton flows

Let f be a non-constant function of the form

$$f(z) = r_1(z)\exp(r_2(z)),$$

where $r_1(z)$ and $r_2(z)$ are rational. As in the case of rational functions, the corresponding Newton flow can be extended as a (real analytic) vector field on the Riemannian sphere S^2, with finitely many equilibria (cf. [A-20]). Now, the questions are: What can be said about the global properties of such a flow? Is it possible to define graphs (drawn on the sphere) classifying the structurally stable flows? How can such graphs be characterized? In comparison with rational Newton flows there is a severe complication: at poles for $r_2(z)$ the function $f(z)$ attains *isolated essential singularities*. So, f is *not* meromorphic. This gives rize to an additional type of local phase portrait. See Fig. 12, where a computer drawn picture is presented of the Newton flow for the function $f(z) = \exp(\dfrac{2z}{z^2-1})$. Note that for $z = \pm 1$, the flow attains so-called "elliptic sectors".

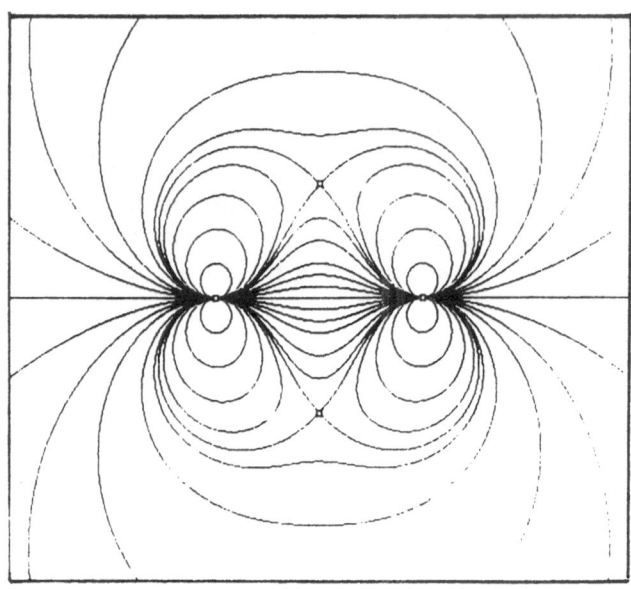

Figure 12

6.4. Continuous versus discrete Newton methods

For meromorphic functions, the relationship between the discrete Newton iteration and the continuous Newton method is extensively studied by Meier [A-21]. The following result has been proved:

Let f be an non-constant meromorphic function and $\check{z}$ a zero for f. We denote (as usual) by $\mathcal{N}(f)$ the Newton flow of f, whereas $\mathcal{N}_h(f)$ denotes the damped Newton iteration $z_{n+1} = z_n - h \cdot f(z_n)/f'(z_n)$, cf. (2). Then, the basins of attraction for $\check{z}$ w.r.t. $\mathcal{N}(f)$ (i.e. the points "moving" to $\check{z}$ along the trajectories of $\mathcal{N}(f)$) will be approximated, for suitably chosen steplength h, by the basins of attraction for $\check{z}$ w.r.t. $\mathcal{N}_h(f)$ (i.e. the "starting" points z_0 for which $\mathcal{N}_h(f)$ tends to $\check{z}$).

Now, let us suppose that $f \in \widetilde{\mathcal{R}}_+$. Then, Meier's results imply the following interesting property of the Newton graph $G(f)$:

When $h \to 0$, then the Julia set with respect to h and f (i.e. the subset of $\mathbf{C}$ for which the above iteration $\mathcal{N}_h(f)$ behaves "badly") converges in a set-theoretical sense to the Newton graph $G(f)$.

Meier's result is based on elements from complex function theory (Koebe's theorem; distortion theory). However, we believe that an independent approach is possible, which merely treats the extension of structurally stable Newton flows to the (compact!) Riemannian sphere S^2. When this idea turns out to be correct, results on the Julia set of the discrete Newton method, similar to those of Meier, will probably be obtained for other classes of complex functions (compare Subsections 6.1 and 6.3). For a first result in this direction (but without treating extended Newton flows) see [A-22].

6.5. Structurally stable Newton flows in general

Up till now, structural stability results for Newton flows are only known in the 2-dimensional case (cf. Remark 4). However, a structural stability analysis in the general case would be very helpfull (especially to understand discrete versus continuous Newton methods). A first step towards such an analysis is the description of the typical local phase portraits of desingularized systems (3). For functions $f : \mathbf{R}^k \to \mathbf{R}^k$, with $k \leq 5$, and under certain regularity conditions, this description is given in [I-5] and [A-23]. In view of these results, a complete classification, similar to the one for rational complex functions, is most unlikely when $k > 2$. Nevertheless, it seems possible to formulate mild conditions under which general desingularized Newton flows behave structurally stable. For a first approach, see [A-24], where some ideas concerning the 3-dimensional case are developed.

Acknowledgement. I would like to thank Dini Heres and Martin Streng for their technical help during the preparation of the manuscript. I am also indebted to an anonymous referee for his constructive review of the first draft of this paper.

References

[I-1]   J. Stoer and R. Bulirsch: *Introduction to numerical analysis*, Springer, New York (1980).

[I-2]   D.G. Luenberger: *Introduction to linear and nonlinear programming*, Addison-Wesley, London (1973).

[I-3]   G.W. Gear: *Numerical initial value problems in ordinary differential equations*, Prentice-Hall, New York (1971).

[I-4]   E. van Groesen: *A kaleidoscopic excursion into numerical calculations of differential equations*, This issue (1992).

[I-5]   H.Th. Jongen, P. Jonker and F. Twilt: *Nonlinear optimization in $\mathbf{R}^n$*, part II, Methoden und Verfahren der Mathematischen Physik, **32**, Peter Lang, Frankfurt a M., Bern, New York (1986).

[I-6]   A.I. Markushevich: *Theory of functions of a complex variable*, Vol. II, Englewood Cliffs, Prentice Hall (1965).

[I-7]   S. Lefschetz: *Differential equations: Geometric theory*, Interscience Publ.

[I-8]   A.A. Andronov, E.L. Leontovich, I.I. Gordon and A.G. Maier: *Theory of bifurcations of dynamical systems on a plane*, Wiley, New York (1973).

[I-9]   J.W. Milnor: *Topology from a differential viewpoint*, Univ. Press of Virginia (1965).

[I-10] P.J. Giblin: *Graphs, surfaces and homology*, Wiley, New York (1977).

[I-11] O.L. Mangasarian: *Nonlinear programming*, Mc. Graw Hill, New York (1986).

[I-12] L. Mirsky: *Transversal theory*, Acad. Press, New York (1971).

[I-13] A.I. Markushevich: *Theory of functions of a complex variable*, Vol.III, Englewood Cliffs, Prentice Hall (1967).

[A-1] E. Hairer and G. Wanner: *Solving ordinary differential equations, Vol. 2, Stiff and differential-algebraic problems*, Springer Series in Computational Mathematics **14**, Springer, Berlin (1991).

[A-2] D. Braess: *Ueber die Einzugbereiche der Nullstellen von Polynomen beim Newton-Verfahren*, Num. Math. **29**, pp. 123-132 (1977).

[A-3] F.H. Branin: *A widely convergent method for finding multiple solutions of simultaneous nonlinear equations*, IBM J. Res. Develop., pp. 504-522 (1972).

[A-4] I. Diener: *On the global convergence of pathfollowing methods to determine all solutions to a system of nonlinear equations*, Math. Prog. **39**, pp. 181-188 (1987).

[A-5] I. Diener: *Trajectory nets connecting all critical points of a smooth function*, Math. Prog. **36**, pp. 340-353 (1986).

[A-6] C.B. Garcia and F.J. Gould: *Relation between several pathfollowing algorithms and local and global Newton methods*, SIAM Review **22**, no. 3, pp. 263-274 (1980).

[A-7] M.W. Hirsch and S. Smale: *Algorithms for solving $f(x) = 0$*, Comm. Pure Appl. Math. **32**, pp. 281-312, (1979).

[A-8] H.Th. Jongen, P. Jonker and F. Twilt: *The continuous Newton method for meromorphic functions*, In: Geometric approaches to differential equations (R. Martini, ed.), Lect. Notes in Math., **810**, Springer, pp. 181-239 (1980).

[A-9] H.Th. Jongen, P. Jonker and F. Twilt: *The continuous, desingularized Newton method for meromorphic functions*, Acta Appl. Math. **13**, pp. 81-121 (1988).

[A-10] H.Th. Jongen, P. Jonker and F. Twilt: *On the classification of plane graphs representing structurally stable rational Newton flows*, Journal of Comb. Theor. Series B, **15**-2, pp. 256-270 (1991).

[A-11] H.B. Keller: *Global homotopies and Newton methods*, In: Recent advances in numerical analysis (C. de Boor, G.H. Golub, eds.), Acad. Press, pp. 73-74 (1978).

[A-12] M. Shub, D. Tischler and R.F. Williams: *The Newtonian graph of a complex polynomial*, SIAM J. Math. Anal., **19**-1, pp. 246-256 (1988).

[A-13] S. Smale: *A convergent process of price adjustment and global Newton methods*, J. Math. Economics **3**, pp. 107-120 (1976).

[A-14] S. Smale: *On the efficiency of algorithms of analysis*, Bull. Am. Math. Soc., **13**-2, pp. 87-121 (1985).

[A-15] M.C. Peixoto: *Structural stability on two dimensional manifolds*, Topology **1**, pp. 101-120 (1962).

[A-16] M.C. Peixoto: *On the classification of flows on 2-manifolds*, In: Dynamical Systems (M.M. Peixoto, ed.), Acad. Press, New York, pp. 389-419 (1973).

[A-17] F. Harary, G. Prins and W.T. Tutte: *The number of plane trees*, Indag. Math. **26** (1964).

[A-18] G.F. Helminck, M. Streng and F. Twilt: *The qualitative behaviour of Newton flows for Weierstrass $\mathcal{P}$-functions*, In preparation.

[A-19] F. Twilt, P. Jonker and M. Streng: *Gradient Newton flows for complex polynomials*, In: Proc. Fifth French-German Conference on Optimization, Lect. Notes in Math. **1405** (S. Dolecki, ed.), pp. 177-190 (1989).

[A-20] W. Wesselink: *De continue Newton methode voor functies van de vorm $R_1(z)\exp(R_2(z))$ met $R_1(z)$ en $R_2(z)$ rationaal*, D-report, Dept. of Appl. Math., Univ. of Twente (1991).

[A-21] H.G. Meier: *Diskrete und kontinuierliche Newton Systeme im Komplexen*, Ph.D. thesis, RWTH Aachen (1991)

[A-22] W. Bergweiler et al.: *Newton's method for meromorphic functions*, Preprint **31**, Lehrstuhl C für Mathematik, RWTH Aachen (1991).

[A-23] H.Th. Jongen, P. Jonker and F. Twilt: *A note on Branin's method for finding the critical points of smooth mappings*, In: Parametric optimization and related topics, (J. Guddat, H.Th. Jongen, B. Kummer and F. Nožička, eds.), Akad. Verlag, Berlin (1987).

[A-24] M. de Graaf: *The continuous Newton method in three dimensions*, D-report, Dept. of Appl. Math., Univ. of Twente (1990).

Recurrence and induction in computer science

A.J. van Zanten
Faculty of Technical Mathematics and Informatics
Delft University of Technology
P.O. Box 5031, 2600 GA Delft, The Netherlands

Abstract: Recurrence relations, or more generally, recursiveness, and proofs by mathematical induction are important ingredients of discrete mathematics. In this introductory paper it will be shown that these concepts can be applied extremely well to computer science. Moreover, the various examples by which these applications are illustrated will also demonstrate that both concepts are intimately related, and that there is a great deal of interaction between them in practical situations.

A. van der Burgh and J. Simonis (eds.), Topics in Engineering Mathematics, 239–265.
© 1992 *Kluwer Academic Publishers.*

1. INTRODUCTION

Combinatorial issues are at the heart of computer science and modern discrete mathematics. The simple question of how many objects exist having a certain property can quickly develop into challenging problems about algorithm analysis, or about the structural symmetry of certain combinatorial objects or of some digital hardware. In this paper we shall focus our attention on *recursion* in the sense of *recurrence relations* and *recursive algorithms*.

Recurrence relations crop up in all places where one can express some variable labelled by one or more discrete parameters (usually called n, m, etc.) in terms of the same variable, but with lower values for its parameters $(n - 1, n - 2, \cdots, m - 1, m - 2, \cdots)$. The same holds for recursively defined objects such as combinations, permutations, or words over a certain alphabet, which satisfy special conditions. Both, variables in a recurrence relation and recursively defined objects have to be defined in a normal, i.e. non-recursive, way for one or more initial values of the parameters. These non-recursive rules are called *initial conditions* or *boundary conditions*.

The fact that variables labelled by discrete parameters play such an important role in computer science, contrary to variables labelled by continous parameters which dominate the more conventional disciplines like physics, is due to the "discrete" nature of computing machinery (bits, bytes, memory places), and to the form of communication between man and machine (characters, words, strings, programs, formal languages).

The concept of *recursive algorithm* or *recursive procedure* (these two terms will be used interchangeably) is well-known from higher level programming languages such as PASCAL and ALGOL. It occurs as soon as in the body of a procedure a statement is used containing a call for the same procedure, but with a lower value for its formal parameter(s). The effect is that during its execution such a procedure calls itself repeatedly, each time by a lower parameter value. To prevent this calling from going on forever ("eternal loop") one has to take care that for some intial value of the parameter(s) the procedure is defined in a non-recursive way.

Another element which frequently arises in discrete mathematics is the *principle of mathematical induction*. This principle has been a well-known feature in other branches of mathematics for a long time, and is mostly applied to prove theorems depending on a discrete variable, almost always a natural number called n.

First the theorem is proved for some smallest value n_0 of n; this is called the *basis*. Then the theorem is proved for $n > n_0$, assuming that it has already been proved for $n - 1$; this is called the *induction*. The attractive thing about such a proof is that it provides infinitely many results with only a finite amount of work.

All three concepts, i.e. recurrence relations, recursive procedures and induction proofs, have recursion (which literally means walking back) as an underlying principle. Moreover, the initial conditions of a recurrence relation, the non-recursive part of a recursive procedure, and the basis of an induction proof all play the same role.

What we want to show is that these three basic elements of discrete mathematics go together extremely well in computer science. For example, the total number of recursively defined objects can often be calculated by solving a recurrence relation. Properties concerning the specific form of these objects, or their relative order are often proved by mathematical induction. The correctness of a recursive algorithm can often be proved by induction as well. The analysis of such an algorithm, or of a computer program based on it, also gives rise to recurrence relations. As an example we mention the complexity of an algorithm, which is the number of times that a specific statement, characteristic for that algorithm, is carried out. This number is often determined by recursive methods.

By means of some concrete examples we shall illustrate the general ideas described above. These are almost all textbook examples, and can also be found in the references mentioned at the end of this article.

2. STRINGS OF SYMBOLS

2.1. BIT STRINGS *

Our first example concerns *bit strings*, i.e. sequences of zeros and ones. One can also speak of *binary numbers* or of *words over the alphabet* {0,1}. We ask for the total number b_n of bit strings of length n which do not have two or more *consecutive* 1-bits. For the sake of convenience we shall call such strings "good strings". Bit strings which contain somewhere the pattern 11 are called "bad strings". To derive a recurrence relation for the numbers $b_n, n = 0, 1, 2, ...$, we observe that we can divide the set of good strings of length n into two subsets: those that have a 0-bit at the right end of the sequence and those that have a 1-bit at the right end. In the first case the 0-bit can be preceded by any good string of length $n - 1$. In the second case we know for sure that the last but one bit is equal to 0, since otherwise we would have a bad string. The pattern 01 at the right end of such a string then can be preceded by any good string of length $n - 2$. Symbolically we represent both types of good strings by the following pictures.

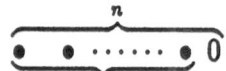

good string of length $n - 1$

good string of length $n - 2$

By definition there are b_{n-1} good strings of the first type, and b_{n-2} of the second type. Hence, we conclude that

$$b_n = b_{n-1} + b_{n-2}, \quad n \geq 0. \tag{1}$$

As soon as we know two numbers b_i and b_{i+1} for some i we can compute all b_j with a higher index, by applying (1) repeatedly. Now, we have $b_0 = 1$, since there is only one good string of length 0: the empty string. Furthermore,

---

0* cf.[3]

$b_1 = 2$, since 0 and 1 are both good strings of length 1. Therefore the initial conditions are

$$b_0 = 1, \quad b_1 = 2. \tag{2}$$

Applying eq. (1) yields $b_2 = 1 + 2 = 3$, $b_3 = 2 + 3 = 5$, $b_4 = 3 + 5 = 8$, etc. (If one feels uneasy about the $n = 0$-case one can remain at the safe side and take $b_1 = 2$ and $b_2 = 3$ as initial conditions; $00, 10$ and 01 are the three good strings of length 2.) Eq.(1) is an example of a *linear recurrence relation of order 2*. In stead of applying (1) several times to compute b_n for a certain value of n, we prefer of course a closed expression for b_n as function of n. There are well-known elegant methods to solve linear recurrence relations. We shall not discuss these methods in general, but we shall illustrate them by solving (1) and similar recurrence relations in the next sections.

Substitution of $b_n = \alpha^n$ in (1) provides us with $\alpha^n = \alpha^{n-1} + \alpha^{n-2}$, or equivalently

$$\alpha^2 - \alpha - 1 = 0. \tag{3}$$

This second order polynomial equation is called *the characteristic equation* of (1). It has two roots, $\alpha_1 = \frac{1}{2}(1 + \sqrt{5})$ and $\alpha_2 = \frac{1}{2}(1 - \sqrt{5})$, and therefore $b_n = \alpha_1^n$ and $b_n = \alpha_2^n$ are both solutions of (1). The theory of linear recurrence relations now tells us that the most general solution of (1) is

$$b_n = \lambda \alpha_1^n + \mu \alpha_2^n = \lambda \left(\frac{1 + \sqrt{5}}{2} \right)^n + \mu \left(\frac{1 - \sqrt{5}}{2} \right)^n. \tag{4}$$

The two constants λ and μ can be determined by requiring $b_0 = 1$ and $b_1 = 2$. We find, doing some elementary calculations,

$$\lambda = \frac{1}{2} \left(1 + \frac{3}{\sqrt{5}} \right), \quad \mu = \frac{1}{2} \left(1 - \frac{3}{\sqrt{5}} \right),$$

and finally

$$b_n = \frac{1}{\sqrt{5}} \left(\frac{1 + \sqrt{5}}{2} \right)^{n+2} - \frac{1}{\sqrt{5}} \left(\frac{1 - \sqrt{5}}{2} \right)^{n+2}. \tag{5}$$

The numbers b_n are called *Fibonacci-numbers*, or more precisely, numbers of Fibonacci-type, since the original Fibonacci-numbers were defined with initial conditions $b_0 = 1$, $b_1 = 1$. Expression (5) can be used for all kinds of properties of these numbers. Here, we only mention the asymptotic behaviour of b_n. Since $\left|\frac{1-\sqrt{5}}{2}\right| < 1$, we infer that for large values of n we have

$$b_n \sim \frac{1}{\sqrt{5}} \left(\frac{1+\sqrt{5}}{2}\right)^{n+2}.$$

2.2. ARITHMETIC EXPRESSIONS *

Our second problem stems from the field of *formal languages*. In many of these languages the concept *arithmetic expression* is defined. We consider arithmetic expressions (abbreviated by *arexp* in this section) without parentheses that are made up of the digits $0, 1, \cdots, 9$ and the binary operation symbols $+, *$ and $/$.

Together these thirteen symbols are the basic symbols for arexps, the formal definition of which is as follows.

digit ::= $0, 1, 2, 3, 4, 5, 6, 7, 8, 9$
operator ::= $+, *, /$
number ::= digit | digit number
arexp ::= number operator number | arexp operator number

In this definition a comma is always a separation sign and the symbol | stands for "or"; both, and | do not belong to the language. We remark that the language elements number and arexp are recursively defined. For the evaluation of arexps there is a hierarchy of operators: multiplication and division are performed before addition. Operations of the same level are performed in their order of appearance as the arexp is scanned from left to right. However, we shall not bother about evaluation here. To prevent dividing by zero we add another rule which we describe, for our convenience, in a less formal way: the pair of basic symbols / 0 is forbidden. One can easily verify that expressions like $2, 05, 2*3, 23, 2+3 / 5+0, 2*000$ are legal arexps. Not allowed are expressions like $+8, 5/0, 2/0 + 1, 7 * +9$. Neither is $7/05$ a legal arexp, although in "every day life" we would accept this as a well defined division of 7 by 5.

Let a_n be the total number of arexps that are made up of n symbols. Then $a_1 = 10$, since the arexps of 1 symbol are the 10 digits.

0* cf.[2]

Furthermore, $a_2 = 100$, accounting for the arexps 00 , 01 , $\cdots$, 99. For $n \geq 3$ we distinguish between two types of arexps. Either the arexp has at its right end a pair of digits, or it has an operator followed by a digit at that place. Symbolically we have the types.

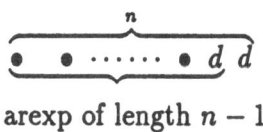

arexp of length $n - 1$

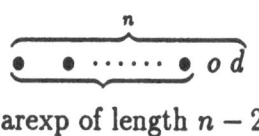

arexp of length $n - 2$

The first case occurs if and only if the first $n - 1$ symbols constitute an arexp of length $n - 1$. Hence, there are $10a_{n-1}$ arexps of this type. The second case occurs if and only if the first $n - 2$ symbols constitute an arexp. Since / 0 is the only operator-digit pair which is forbidden, there are $29a_{n-2}$ arexps of the second type.

Therefore, the recurrence relation and initial conditions respectively are

$$a_n = 10a_{n-1} + 29a_{n-2}, \qquad n > 2, \tag{6}$$

$$a_1 = 10 , \ a_2 = 100. \tag{7}$$

Substitution of $a_n = \alpha^n$ in (6) gives the characteristic equation

$$\alpha^2 - 10\alpha - 29 = 0 \quad , \tag{8}$$

which has solutions $\alpha_{1,2} = 5 \pm 3\sqrt{6}$. So it follows that

$$a_n = \lambda(5 + 3\sqrt{6})^n + \mu(5 - 3\sqrt{6})^n,$$

where λ en μ have to be chosen such that the initial conditions (7) are satisfied. After a simple calculation we find

$$a_n = \frac{5}{3\sqrt{6}} \left[\left(5 + 3\sqrt{6}\right)^n - (5 - 3\sqrt{6})^n \right], \qquad n > 0. \tag{9}$$

2.3. BIT STRINGS WITH A SPECIAL PATTERN ON THE LAST POSITION *

Again we consider bit strings of length n. A *pattern* consists of one or more consecutive zeros and ones, like 01 or 1101. A pattern is said to occur at the kth position, if, in scanning the string from *left to right*, the last bit of the pattern is on the kth position of the string. As an example we take the pattern 010 and the bit string

$$110101010101$$

According to our definition the pattern occurs at the fifth position. We now give a supplementary rule in order to define precisely the number of times that a pattern occurs in a string. Two patterns (of the same type) are not allowed to overlap. Thus after a pattern occurs, scanning starts all over again to search for the next occurrence of that pattern in the string. In our example 010 occurs at the 5th and at the 9th position, but *not* on the 7th and the 11th position.

Let c_n be the number of bit strings of length n such that 010 occurs at the nth position. To derive a recurrence relation we consider *all* strings of length n with 010 at the last three positions

The total number of these strings is 2^{n-3}. This set of strings can be divided into two disjunct subsets:

(i) strings with 010 on the nth position;
(ii) strings not with 010 on the nth position.

By definition there are c_n strings of type (i). The only reason that a string is of type (ii) is that 010 occurs at the $(n-2)$th position, which prevents the last three bits from being acknowledged as an occurrence of the pattern at the nth position. Hence, in case of type (ii) the string is of the form

---

0* cf.[3]

and so there are c_{n-2} strings of this type. Consequently we have

$$c_n + c_{n-2} = 2^{n-3}, \tag{10}$$

$$c_1 = 0, \qquad c_2 = 0. \tag{11}$$

The numbers c_1 and c_2 are equal to 0, since strings of length < 3 cannot contain a pattern of length 3.

Eq. (10) is a linear recurrence relation of the first order. However, there is a difference with eqs. (1) and (6). Contrary to these equations, which are *homogeneous*, eq.(10) is *inhomogeneous* because of the *rhs* being not zero. From the theory of linear recurrence relations we know that the general solution of (10) is the sum of an arbitrary solution (called a *particular solution*) and the general solution of the homogeneous equation

$$c_n + c_{n-2} = 0. \tag{12}$$

Substituting $c_n = \alpha^n$ gives the characteristic equation

$$\alpha^2 + 1 = 0, \tag{13}$$

with characteristic roots $\alpha_{1,2} = \pm i$. For the general solution of (12) we can write

$$c_n = \lambda i^n + \mu(-i)^n = (\lambda + \mu)\cos\frac{n\pi}{2} + i(\lambda - \mu)\sin\frac{n\pi}{2} =$$

$$\sigma\cos\frac{n\pi}{2} + \tau\sin\frac{n\pi}{2}. \tag{14}$$

The constants λ and μ in (14) are arbitrary complex numbers. Since we need a real solution we chose them such that σ and τ are both real. A particular solution of (10) can be found by substitution of $c_n = A2^n$, giving $A2^n + A2^{n-2} = 2^{n-3}$, and hence $A = \frac{1}{10}$. We conclude that the general solution of (10) can be written as $c_n = \frac{1}{10}2^n + \sigma\cos\frac{n\pi}{2} + \tau\sin\frac{n\pi}{2}$. Finally the real arbitrary constants σ and τ are determined by the initial conditions (11), yielding

$$c_n = \frac{1}{10}\left[2^n - \cos\frac{n\pi}{2} - 2\sin\frac{n\pi}{2}\right], \qquad n \geq 0. \tag{15}$$

3. GRAY CODES AND RECURSIVENESS

3.1. THE BINARY REFLECTED GRAY CODE *

In the figure below we present two tables. In both tables all bit strings of length 3 are listed, or, equivalently, the integers $0, 1, \cdots, 7$ are listed in their binary representation.

$$
\begin{array}{cc}
000 & 000 \\
001 & 001 \\
010 & 011 \\
011 & 010 \\
100 & 110 \\
101 & 111 \\
110 & 101 \\
111 & 100 \\
\end{array}
$$

Fig. 3.1

In the left table the strings are ordered lexicographically, or, in terms of binary numbers, they are arranged according to increasing values. If we go from one string to the next the number of bits to be changed varies. For example, if we go from 000 to 001 there is only one bit which changes. If however we go from 001 to 010 there are two such bits, and if we go from 011 to 100 there are even three of them.

In the right table the same strings are listed, but in different order. If we compare any pair of successive strings, there is always *only one bit* which changes. Such sequences of bit strings are known as *Gray codes*. More precisely, an n-bit Gray code is an ordered sequence of the 2^n n-bit strings, called *codewords*, such that successive codewords differ by the change of a single bit. We shall loosely speak of a Gray-ordered list. If the first and the last word also differ in one bit one speaks of a cyclic Gray code.

These codes play a role in various parts of computer mathematics and are applied for various reasons. Originally they were developed for the purpose of minimizing the number of erroneous bits in the transmission of bit strings

---

[0]* cf.[6]

as analog signals. If bit strings are coded arithmetically, i.e. such that a bit string is represented by the decimal number to which it corresponds, and if that number is transmitted in some way, then an error of one unit in the analog signal can cause a large number of bit errors if one uses a lexicographic code (cf. Fig. 3.1, left column). However, if one uses a Gray code an error of one in the analog signal causes only one bit error.

There are many n-bit Gray codes, but we will be concerned only with the binary-reflected Gray code (called *the* Gray code from now on). This code is defined recursively. Let $G(n)$ be a Gray code written as a $2^n \times n$-matrix of codewords g_i^n, $0 \le i < 2^n$,

$$G(n) = \begin{bmatrix} g_0^n \\ g_1^n \\ \vdots \\ g_{2^n-1}^n \end{bmatrix}. \tag{16}$$

Then the $2^{n+1} \times (n+1)$-matrix

$$G(n+1) := \begin{bmatrix} 0G(n) \\ 1G(n)^R \end{bmatrix} = \begin{bmatrix} 0g_0^n \\ \vdots \\ 0g_{2^n-1}^n \\ 1g_{2^n-1}^n \\ \vdots \\ 1g_0^n \end{bmatrix} \tag{17}$$

is also a Gray code, as one can verify immediately by *induction*. The notation $G(n)^R$ stands for the matrix $G(n)$, but with its rows in reversed order. Since $G(n)$ is a Gray code we have that $G(n)^R$ is also a Gray code. Therefore successive words in the first half, $0G(n)$, of $G(n+1)$ differ in precisely one bit, and the same holds for the second half, $1G(n)^R$. The last word of $0G(n)$ and the first word of $1G(n)^R$ differ only in the front bit. Starting with the trivial Gray code

$$G(1) = \begin{bmatrix} 0 \\ 1 \end{bmatrix}, \tag{18}$$

we obtain by (17) a Gray code for all $n \geq 1$, which is even cyclic. From eqs. (17) and (18) we infer

$$G(2) = \begin{bmatrix} 00 \\ 01 \\ 11 \\ 10 \end{bmatrix}$$

and, by applying once more eq. (17), we get $G(3)$, which is the right table in Fig. 3.1.

3.2. TRANSITION SEQUENCES *

Next we want to generate the codewords of $G(n)$ for some fixed integer n, without using (17). As an auxiliary tool we introduce the notion of *transition sequence*. We number the columns of $G(n)$ by the numbers $0, 1, \cdots, n-1$, from *right to left*. If t_i is the index of the bit which changes in the transition from g_i^n to $g_{i+1}^n, 0 \leq i \leq 2^n - 2$, then the transition sequence is defined as

$$T_n := (t_0, t_1, \cdots, t_{2^n-2}). \tag{19}$$

For example, $G(3)$ has the transition sequence $T_3 = (0, 1, 0, 2, 0, 1, 0)$. The general form of T_n is determined by the following recurrence relation and initial condition

$$T_{n+1} = (T_n, n, T_n), \quad n > 0, \tag{20}$$

$$T_1 = (0). \tag{21}$$

This needs a short proof. By the recursive definition (17) we have $T_{n+1} = (T_n, n, T_n^R)$, $n > 0$. Since $T_1^R = T_1$, it follows that $T_n^R = T_n$ and consequently relation (20) holds for all $n > 0$.

In stead of eqs. (17) and (18) we can use an alternative definition for $G(n)$

0* cf.[6]

$$G(1) = \begin{bmatrix} 0 \\ 1 \end{bmatrix}, \quad G(n+1) = \begin{bmatrix} g_0^n 0 \\ g_0^n 1 \\ g_1^n 1 \\ g_1^n 0 \\ \vdots \end{bmatrix}. \tag{22}$$

It is obvious by induction that the code defined by (22) is a Gray code. Its equivalence to the code of eqs. (17) and (18) can also be proved by induction. We omit such a proof here. It follows from (22) that, if T_n is as given by (19), the sequence T_{n+1} can be written as

$$T_{n+1} = (0, t_0 + 1, 0, t_1 + 1, ..., 0, t_{2^n-2} + 1, 0). \tag{23}$$

For $n = 2$ one can easily verify that relation (23) is true.

If the transition sequence T_n is known for a certain value of n, it is possible, starting with $g_0^n = 00 \cdots 0$, to produce the code $G(n)$, without making use of a recursive definition. In Section 3.4 we shall discuss an algorithm based on this idea.

3.3. INDEXING THE GRAY CODE *

An obvious question is how to construct a codeword $g_i \in G(n)$, *without* producing all preceding codewords. The answer is surprisingly simple. Let

$$i = (b_{n-1} b_{n-2} \cdots b_0)_2 \tag{24}$$

be the binary representation of the index i. Then the bits $g_{i,k}^n$ of g_i^n can be obtained by the rule

$$g_{i,k}^n = b_{k+1} + b_k \pmod{2}, \tag{25}$$

for $0 \le k \le n - 1$. The bit b_n, which we need in (25) for $k = n - 1$, is omitted in representation (24), since it is always equal to 0. A more practical formulation of (25) is as follows

---

0* cf.[6]

$$
\begin{array}{rcccccc}
i & : & b_{n-1} & b_{n-2} & \cdots & b_1 & b_0 \\
\lfloor \frac{i}{2} \rfloor & : & 0 & b_{n-1} & \cdots & b_2 & b_1 \\
& & - & - & - & - & - \quad \oplus \\
g_i^n & : & g_{i,n-1}^n & g_{i,n-2}^n & \cdots & g_{i,1}^n & g_{i,0}^n
\end{array}
\tag{26}
$$

where $\oplus$ means bitwise addition modulo 2. For example, the codeword g_6^3 of $G(3)$ is obtained by

$$
\begin{array}{rccc}
6 & : & 1 \quad 1 \quad 0 \\
3 & : & 0 \quad 1 \quad 1 \\
& & - \quad - \quad - \quad \oplus \\
g_6^3 & : & 1 \quad 0 \quad 1
\end{array}
$$

The rules (25) and (26) can again be proved by induction. Assume that (26) holds for a certain value of $n \geq 1$. Next we want to determine $g_i^{n+1} \in G(n+1)$, for a given index i. For $0 \leq i \leq 2^n - 1$ we can write $i = (0b_{n-1}b_{n-2}\cdots b_0)_2$, and for $2^n \leq i \leq 2^{n+1} - 1$ we have $i = (1b_{n-1}b_{n-2}\cdots b_0)_2$. In the first case rule (26) provides us with

$$
\begin{array}{rccccccc}
i & : & 0 & b_{n-1} & b_{n-2} & \cdots & b_1 & b_0 \\
\lfloor \frac{i}{2} \rfloor & : & 0 & 0 & b_{n-1} & \cdots & b_2 & b_1 \\
& & - & - & - & - & - & - \quad \oplus \\
g_i^{n+1} & : & 0 & g_{i,n-1}^n & g_{i,n-2}^n & \cdots & g_{i,1}^n & g_{i,0}^n
\end{array}
$$

which is the correct answer, since $0g_i^n$ has the same index in $G(n+1)$ as g_i^n has in $G(n)$ (cf. eq.(17)). In the second case we obtain, by using (26),

$$
\begin{array}{rccccccc}
i & : & 1 & b_{n-1} & b_{n-2} & \cdots & b_1 & b_0 \\
\lfloor \frac{i}{2} \rfloor & : & 0 & 1 & b_{n-1} & \cdots & b_2 & b_1 \\
& & - & - & - & - & - & - \quad \oplus \\
g_i^{n+1} & : & 1 & \bar{g}_{i,n-1}^n & g_{i,n-2}^n & \cdots & g_{i,1}^n & g_{i,0}^n
\end{array}
$$

where $\bar{g}_{i,n-1}^n$ is the complement of $g_{i,n-1}^n$, modulo 2. The position of the codeword $\bar{g}_{i,n-1}^n \, g_{i,n-2}^n \cdots g_{i,0}^n$ in $G(n)$ is the position of $g_{i,n-1}^n \, g_{i,n-2}^n \cdots g_{i,0}^n$ reflected with respect to the middle of the list. Hence, we find again the correct answer (cf. again eq. (17)).

Rule (26) can be written in matrix form

$$
\underline{g} = P\underline{b},
\tag{27}
$$

with the vectors $\underline{b}$ and $\underline{g}$ defined as $\underline{b} := (b_{n-1}, b_{n-2}, \cdots, b_0)$ and $\underline{g} := \left(g_{i,n-1}^n, g_{i,n-2}^n, \cdots, g_{i,0}^n\right)$, and the matrix P as

$$P = \begin{bmatrix} 1 & 0 & 0 & \cdots & 0 \\ 1 & 1 & 0 & \cdots & 0 \\ 0 & 1 & 1 & \cdots & 0 \\ \vdots & \vdots & \vdots & \vdots & \vdots \\ 0 & 0 & 0 & \cdots & 1 \end{bmatrix}. \qquad (28)$$

All additions and multiplications implied by (27) are in the field $GF(2)$, i.e. are carried out modulo 2.

We can easily invert (27), which yields

$$\underline{b} = P^{-1}\underline{g}, \qquad (29)$$

with

$$P^{-1} = \begin{bmatrix} 1 & 0 & 0 & \cdots & 0 \\ 1 & 1 & 0 & \cdots & 0 \\ 1 & 1 & 1 & \cdots & 0 \\ \vdots & \vdots & \vdots & \vdots & \vdots \\ 1 & 1 & 1 & \cdots & 1 \end{bmatrix}, \qquad (30)$$

or equivalently, for $0 \leq j \leq n-1$,

$$b_j = g_{i,n-1}^n + g_{i,n-2}^n + \cdots + g_{i,j}^n \quad (mod\ 2). \qquad (31)$$

Eq. (31) expresses a simple rule how to compute the index i of a given codeword $g_i^n \in G(n)$. As an example we take the word $g_i^3 := 111 \in G(3)$. Using eq.(31) gives us $i = (101)_2 = 5$.

3.5. A MINIMAL-CHANGE ALGORITHM *

The addition of two bits modulo 2 in (26) is the same as the "exlusive or" for two bits. Hence, one could use (26) to convert the indices i to g_i^n, $0 \leq i \leq 2^n - 1$, very rapidly in machine language. However, in a higher-level programming language operations like shifting and an "exclusive or" are inefficient. Therefore we present an algorithm which generates the codewords of $G(n)$ successively, changing each time the only bit which has to be changed

---

0* cf.[6]

to obtain the next codeword. Since changing a single bit is the minimal change possible, such an algorithm is called a *minimal-change algorithm*. It is sufficient to generate the transition sequence T_n, defined in (19), in an efficient way. It appears that this can be done by using a so-called *stack*. This is a data structure which has the nature of a dynamical sequence, i.e. its length and contents vary throughout the execution of the program. In the case of a stack insertion and deletion occur only at one end, called the *top* of the sequence. Correspondingly, the other end is called the *bottom*. In the next algorithm

S is a stack, such that its top element is always equal to the index indicating the bit of the word g_i^n which has to be changed to get the next word g_{i+1}^n. For reasons of convenience we shall denote g_i^n by g, which implies that in the algorithm the variables $g_0, g_1, \cdots, g_{n-1}$ represent the bits of the codeword g to be printed.

$S \leftarrow$ stack ;
$i := 0$;

for j:=n step (-1) to 0 do $\begin{cases} g_j := 0; \\ S \leftarrow j; \end{cases}$

(*) while $i < n$ do $\begin{cases} \text{print } (g_{n-1}g_{n-2}\cdots g_0); \\ i \leftarrow S; \\ g_i := 1 - g_i; \\ \text{for } j := i - 1 \text{ step } (-1) \text{ to } 0 \text{ do } S \leftarrow j \end{cases}$

The stack initially contains $n, n - 1, \cdots, 0$, with n on the bottom and 0 on the top. Each time the while-statement (*) is executed the top element of S is deleted and its contents are given to the variable i (*pop-operation*). Then the bit g_i is changed and the elements $i - 1, i - 2, \cdots, 0$ are pushed onto the stack (*push-operation*). For $i = 0$ no elements are pushed on the stack.

To illustrate the working of the algorithm we present a concrete example. In the case of $n = 3$ we give the values of the variable i, the contents of the stack S, and the value of the word $g := g_2 g_1 g_0$, at the various moments during the execution of the algorithm, when the condition $i < n$ in (*) is tested.

i	S	g
0	3, 2, 1, 0	000
0	3, 2, 1	001
1	3, 2, 0	011
0	3, 2	010
2	3, 1, 0	110
0	3, 1	111
1	3, 0	101
0	3	100
3		

As one can see, in this example the top of S runs through the sequence of values $0, 1, 0, 2, 0, 1, 0, 3$. Apart from the last value these numbers constitute the transition sequence T_3, and therefore, each time, the next codeword of $G(3)$ is printed. Finally, the last value of the top of S is given to i. In the next test the condition $i < n$ then turns out to be false and the algorithm stops.

The above series of events is typical for the general case. We shall prove now by induction the following statement. The top of S contains successively the elements of the sequence (T_n, n) at the moments when the condition $i < n$ in (*) is tested. More specifically, the top contains the elements of T_n, if $i < n$ is true, whereas it contains n, if $i < n$ is false. Hence, for $i < n$ the relevant bit g_i will be changed each time, and for $i = n$ the algorithm stops.

For $n = 1$ we can verify rather easily that the statement is correct. Assume next that the statement is correct for some value $n - 1 (n > 1)$, i.e. the top runs through the elements of $(T_{n-1}, n-1)$, if the algorithm is executed for the value $n - 1$. Then we apply the algorithm for the value n. Initially S contains the elements $n, n-1, \cdots, 0$. Next the top of S runs through $(T_{n-1}, n-1)$ because of the induction assumption. If the top contains $n - 1$ the algorithm does not stop now, since $n - 1 < n$. Hence, $i := n - 1$, $g_{n-1} := 1 - g_{n-1}$ and, after the execution of the for-statement in (*), S contains the elements $n, n - 2, n - 3, \cdots, 0$. The only difference with the initial contents of S is the element n in stead of $n - 1$ at the bottom. Therefore, in the sequel of the execution the top of S takes on the values of the sequence (T_{n-1}, n), again by the induction assumption. After all the top of S runs through the sequence $(T_{n-1}, n - 1, T_{n-1}, n) = (T_n, n)$ (cf. eq.(20)), and hence, the statement is correct for the value n. So we have proved, by induction, the correctness of the algorithm.

4. RECURSIVE ALGORITHMS

The algorithm of Section 3.5 was not recursive, since the algorithm did not invoke itself. In this chapter we shall discuss some really recursive algorithms, and we shall show that recurrence relations and induction are appropriate tools to attack problems triggered by such algorithms. A standard example illustrating these ideas is the well-known *Tower of Hanoi Problem* which we shall discuss in Sections 4.1. and 4.2. Some other algorithms which clearly show the power of recurrence and induction deal with lists of numbers to be sorted to increasing or decreasing values. These *sorting algorithms* are the subject of Sections 4.3 and 4.4.

4.1. THE TOWER OF HANOI PROBLEM *

We are given three pegs A, B and C and n discs of increasing size. These discs are piled up on peg A in increasing size from top to bottom, and are to be transferred to peg C, using peg B as an auxiliary peg. Only *one* disc may be moved at a time, from any peg to any other peg, subject to the condition that *no disc may ever rest above a smaller disc* .

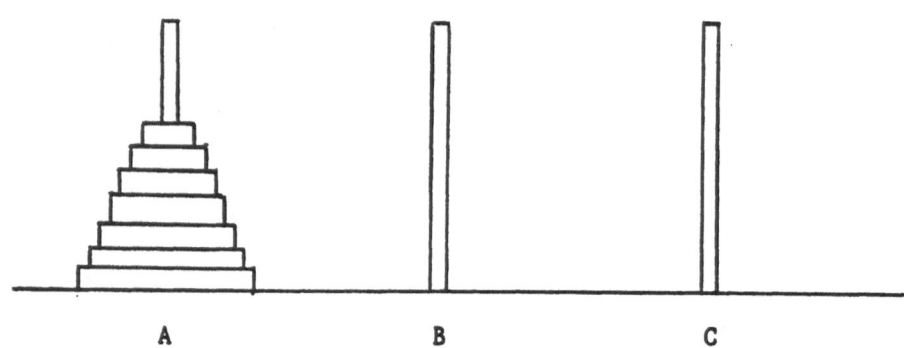

$$A \qquad\qquad\qquad B \qquad\qquad\qquad C$$

0* cf.[1,2,3,6,7]

Proving the existence of a solution for this problem is extremely simple. Assume that there is a solution if there are $n-1$ discs. By applying this solution, we transfer the upper $n-1$ discs on peg A to peg B, using C as an auxiliary peg. We then move the remaining disc from peg A to peg C, and next, again applying the solution for $n-1$ discs, we transfer the discs on peg B to peg C, using A as an auxiliary peg. Since the problem can trivially be solved for $n = 1$, by moving the only disc on peg A directly to peg C, now we have proved by induction that there is a solution for all $n \geq 1$.

Not only did we prove its existence, but we also described the solution in a recursive way. We now present this solution in the form of a recursive procedure. The name of the procedure is HANOI, and it has four parameters n, A, B and C. The first parameter indicates that the procedure transfers n discs, whereas the other three are the names of the initial peg, the auxiliary peg and the final peg respectively.

$$\text{HANOI (n, A, B, C)};$$
$$\textbf{if } n=1 \textbf{ then } \text{move disc from A to C}$$
$$\textbf{else} \begin{cases} \text{HANOI } (n-1, A, C, B); \\ \text{move disc from A to C ;} \\ \text{HANOI } (n-1, B, A, C) \end{cases}$$

It will be clear that this procedure precisely reflects the steps described earlier in the induction proof. At this moment we do not bother about the implementation of the procedure in some computer language. Neither do we distinguish between formal and actual parameters. (A procedure heading HANOI (m, X, Y, Z), and a call for the procedure with actual values n, A, B and C for the parameters would have been nicer.)

In stead we want to answer the question of how many disc moves or *displacements* have to be performed when we apply the above algorithm to transfer n discs. Let a_n be this number. Since the displacement of a disc is the only essential operation to be carried out when executing the algorithm, we define a_n as the *complexity* of HANOI (n, A, B, C) for $n \geq 1$. (Cf. also the Introduction).

From the recursive definition of HANOI we immediately infer that

$$a_n = 2a_{n-1} + 1, \quad n > 1, \tag{32}$$

and

$$a_1 = 1. \tag{33}$$

Eq.(32) is an inhomogeneous, linear recurrence relation of order 1. Its characteristic equation $\alpha - 2 = 0$ has as only characteristic root $\alpha_1 = 2$. Hence, the general solution of the homogeneous equation $a_n - 2a_{n-1} = 0$ is $a_n = \lambda 2^n$. Since $a_n = -1$ is a (particular) solution of (32) we find as the most general solution of eq. (32)

$$a_n = \lambda 2^n - 1 \tag{34}$$

Finally the initial condition (28) yields $\lambda = 1$, and so

$$a_n = 2^n - 1. \tag{35}$$

A faster method to solve eq.(32) would be the substitution $b_n = a_n + 1$, yielding $b_n = 2b_{n-1}$, and $b_1 = 2$, from which we immediately conclude $b_n = 2^n$.

The expression (35) shows us that the complexity a_n grows exponentially as n increases. For not too small values of n the value of a_n is already enormous. For example, we have $a_{30} = 2^{30} - 1 > 10^9$. This inequality implies that a person who dutifully moves discs, day and night, at a rate of one move per second, will need almost his whole lifetime to transfer a tower of 30 discs.

4.2. AN ITERATIVE ALGORITHM DERIVED BY INDUCTION

The algorithm of the previous section is a typical computer algorithm. Human beings are not able very well "to think recursively", at least not four or more levels deep. The person mentioned at the end of Section 4.1 needs a so-called *iterative algorithm*, i.e. a precise prescription of all successive disc moves to be made.

It appears that we can reformulate the HANOI-algorithm in an iterative version. Moreover, a surprising relationship occurs with the Gray codes of the previous chapter. To derive these results we label the n discs to be transferred by the labels $0, 1, \cdots, n - 1$ from small to large. The HANOI-algorithm requires $2^n - 1$ displacements. We label these displacements by the labels $0, 1, \cdots, 2^n - 2$. Suppose that at the ith displacement disc d_i has to be displaced from one peg to another ($d_i \in \{0, 1, \cdots, n-1\}$). Then we can introduce a *displacement sequence* D_n, defined as

$$D_n := (d_0, d_1, \cdots, d_{2^n-2}). \tag{36}$$

By executing HANOI we first have to transfer a tower with discs $0, 1, \cdots, n-2$, then we have to displace disc $n-1$, and finally we have to transfer again a tower with discs $0, 1, \cdots, n-2$. Hence, we have the following recurrence relation

$$D_n = (D_{n-1}, n-1, D_{n-1}), \ n > 1, \tag{37}$$

$$D_1 = (0). \tag{38}$$

If we compare this with relations (20) and (21) we may conclude that D_n is identical to the transition sequence of the Gray code $G(n)$, or

$$D_n = T_n, \quad n \geq 1. \tag{39}$$

At first glance this seems to be a nice and unexpected, but otherwise useless relationship between two different problems. As soon as we know the transition sequence T_n we also know the labels of the discs to be displaced successively. However, it is not clear yet to which peg they have to be moved each time. Moreover, it seems that the recursion of HANOI has been shifted now to the recursive nature of T_n.

Fortunately we also know the expression (23) for T_n, which yields

$$D_n = (0, d_0 + 1, 0, d_1 + 1, 0, \cdots, 0, d_{2^{n-1}-2} + 1, 0), \tag{40}$$

where $(d_0, d_1, \cdots, d_{2^{n-1}-2})$ is the displacement sequence D_{n-1}, $n > 1$. It follows from (40) that we have to displace alternately the smallest disc 0 and a disc with a higher label. The precise values $d_0 + 1, d_1 + 1, \cdots$ of these labels are irrelevant however, since every time there is only one disc, other than disc 0, which can be displaced without violating the rule that a disc is not allowed to be put on a smaller disc. This unique disc is the smaller of the two top discs on the pegs not holding disc 0. Furthermore it can only be moved to one of the other two pegs, because of the same rule.

The only remaining problem is to which peg disc 0 has to be displaced every time. The answer appears to be simple. If n is even, disc 0 has to be displaced such that, during the execution of the algorithm, it visits the three pegs according to $A, B, C, A, B, C, A, \cdots$ This way of moving will be denoted by the cyclic permutation (A, B, C). If n is odd, disc 0 has to be displaced all the time according to the cyclic permutation (A, C, B).

The proof is again by induction. Let $n > 1$ and assume that the rules are true for $n - 1$. Consider HANOI (n, A, B, C), and let n be odd. During its execution first HANOI $(n - 1, A, C, B,)$ is carried out. Due to the induction assumption disc 0 then moves according to the permutation (A, C, B), since $n - 1$ is even in this case. After having displaced the largest disc from A to C, the procedure HANOI $(n - 1, B, A, C)$ is carried out. Now, by the induction assumption, disc 0 moves according to the permutation (B, A, C) $= (A, C, B)$. Hence, throughout the entire execution disc 0 moves according to (A, C, B).

If n is even we can prove analogously that disc 0 moves according to (A, B, C).

Since the rules are trivially true for $n = 1$, we now have proved the rules for all $n \geq 1$.

After all we have derived an extremely simple iterative algorithm. The only thing we have to memorize is the cyclic direction of disc 0. Between any two consecutive moves of disc 0 we have to perform the only displacement possible for discs with a higher label.

4.3. QUICKSORT *

We consider the problem of *sorting* a list, or sequence, of n elements $x_1, x_2, ..., x_n$. We assume that any pair $\{x_i, x_j\}$ of these elements can be compared with each other, and that either $x_i < x_j$ or $x_j < x_i$. The total ordering of such a list, i.e. the rearranging of the elements according to increasing values, can be obtained by comparing all pairs of distinct elements. There are $\frac{1}{2}n(n-1)$ pairs, and so $\frac{1}{2}n(n-1)$ is an *upper bound* on the number of comparisons. In most cases this number is the determining factor in the performance of a sorting algorithm, since it predominates over the number of other actions. Therefore we define the *complexity* of a sorting algorithm as the number of comparisons it requires. However, this number depends on the input, since the list to be sorted could have been ordered already partially. Therefore we can speak of a *worst-case-complexity* of a sorting

---

[0]* cf.[4,6,7]

algorithm, which is the number of comparisons for the worst-case input. On the other hand we have the *average-case-complexity*, which is the number of comparisons averaged over all input sequences. Both types of complexities are a good measure for the efficiency of an algorithm.

The sorting algorithm QUICKSORT works as follows. An element a of the list to be sorted is selected randomly. If there is no other element in the list, then the ordering is completed. Otherwise its position in the ordering is determined by comparing it with all other elements in the list. This requires $n-1$ comparisons. The remainder of the list is now divided into two sublists, one containing the elements less than a, the other all elements greater than a. These sublists have at most $n-1$ elements each, and can be sorted recursively.

procedure QUICKSORT (S);
if $|S| = 1$ **then** print (S)

$$
\textbf{else} \begin{cases}
\text{choose } a \text{ randomly from } S; \\
S_1 := \text{ list of elements } x \text{ with } x \in S \text{ and } x < a; \\
S_2 := \text{ list of elements } x \text{ with } x \in S \text{ and } x > a; \\
\text{QUICKSORT } (S_1); \\
\text{print } (a); \\
\text{QUICKSORT } (S_2)
\end{cases}
$$

We shall now prove that the average-case-complexity is asymptotically equal to $O(n \; log \; n)$

Let t_n be the average-case-complexity of QUICKSORT (S), if $|S| = n > 0$. Then clearly

$$t_1 = 0 , \quad t_2 = 1. \tag{41}$$

If the randomly chosen element a in the body of QUICKSORT has position r when S has been sorted, then $|S_1| = r - 1$ and $|S_2| = n - r$. Assuming that every choice of a, from position 1 to n in the sorted list, is equally likely, we can write

$$t_n = n - 1 + \frac{1}{n} \sum_{r=1}^{n} (t_{r-1} + t_{n-r}). \tag{42}$$

Here we applied the recursive nature of QUICKSORT. Rewriting the rhs of (42) gives for $n > 0$

$$t_n = n - 1 + \frac{1}{n}\sum_{r=1}^{n-1} t_r + \frac{1}{n}\sum_{r=1}^{n} t_{n-r}$$

$$= n - 1 + \frac{2}{n}\sum_{r=1}^{n-1} t_r. \qquad (43)$$

To solve this recurrence relation we introduce

$$x_n = \sum_{r=1}^{n-1} t_r \ , \ n > 0, \qquad (44)$$

which implies

$$t_n = x_{n+1} - x_n. \qquad (45)$$

Substitution of (44) and (45) in (43) gives

$$x_{n+1} - \frac{n+2}{n}x_n = n - 1 \ , n > 0. \qquad (46)$$

Notice that eq. (46) is linear, but that the coefficients of the x-variables are not all constants with respect to n. To overcome this difficulty we define another auxiliary variable for $n > 0$

$$y_n = \frac{x_n}{n(n+1)}, \qquad (47)$$

which finally yields the equation

$$y_{n+1} - y_n = \frac{3}{n+2} - \frac{2}{n+1} \ , n > 0. \qquad (48)$$

One could solve (48) by applying the general theory for linear equations with constant coefficients, as was done in Sections 2 and 3. We shall turn to a different approach however. Addition of the eqs. (48) for $n = 1, 2, 3, \cdots, k-1$ yields for $k > 1$

$$y_k - y_1 = \sum_{r=2}^{k}\left(\frac{3}{r+1} - \frac{2}{r}\right) = \frac{3}{k+1} - 1 + \sum_{r=3}^{k}\frac{1}{r}.$$

Substituting $k = n$ and applying (45), (47) and the initial condition $y_1 = x_1 = 0$ now provides us with the expression

$$t_n = -4n + 2(n+1) \sum_{r=1}^{n} \frac{1}{r}. \tag{49}$$

From calculus we know that the *harmonic number* $H_n := \sum_{r=1}^{n} \frac{1}{r}$ asymptotically behaves like

$$\lim_{n \to \infty} (H_n - \ell n \ n) = \gamma, \tag{50}$$

where γ is called *Euler's constant*, with value $\gamma = 0.5772.....$.
It follows from eqs. (49) and (50) that

$$t_n = O(n \ \ell n \ n) \tag{51}$$

For large values of n QUICKSORT takes on average $O(n \ \ell n \ n)$ comparisons.

4.4. MERGESORT *

This sorting algorithm is similar to QUICKSORT. It divides the sequence to be sorted into two equal subsequences or, if n is odd, into two nearly equal subsequences. After sorting them recursively it "merges" the two subsequences. This is done as follows. Let $A := a_1, a_2, \cdots, a_r$ and $B := b_1, b_2, \cdots, b_s$ be the sorted subsequences. First a_1 and b_1 are compared, and the smaller - a_1 say - is deleted and placed in the first position of the sorted sequence. Next a_2 and b_1 are treated similarly. At any stage the two first elements of the remaining subsequences are compared, until one of the sequences is exhausted. Finally the remainder of the longer sequence is added to the sorted sequence. It will be clear that the merging of A and B will take *at most* $r + s - 1$ comparisons.

We now present an algorithm which sorts according to the above approach a sequence $x_1, x_2, \cdots, x_n$. In this algorithm a procedure MERGE (S,T) is used, which takes two sorted sequences S en T as input and produces as output a sequence consisting of the elements of S and T in sorted order. Since S and T are themselves sorted, MERGE requires at most $|S| + |T| - 1$ comparisons, as was argued before. For the sake of simplicity we do not

---

[0]* cf.[4,6,7]

define MERGE explicitly. The recursive procedure SORT (i,j), to be defined
below, sorts the subsequence $x_i, x_{i+1}, \cdots, x_j$.

> **procedure SORT (i,j);**
> **if $i = j$ then print (x_i)**
>
> $\qquad$ else $\begin{cases} m := \lfloor (i+j)/2 \rfloor; \\ MERGE(SORT(i,m), SORT(m+1,j)) \end{cases}$

In order to sort the sequence $x_1, x_2, \cdots, x_n$ we call SORT (1,n).

We shall prove that the worst-case-complexity of SORT (1,n) is asymptotically of the order $O(n \ln n)$. Let s_n be the *greatest number* of comparisons required to sort a sequence of length n when using SORT. Then obviously we have the recurrence relations

$$\begin{cases} s_{2n} = 2s_n + 2n - 1 \\ s_{2n+1} = s_{n+1} + s_n + 2n \end{cases} \tag{52}$$

with the initial condition

$$s_1 = 0. \tag{53}$$

By mathematical induction we now show that,

$$s_n \leq (k-1)2^k + 1, \tag{54}$$

for $n \leq 2^k$.

For $n = 1$ this is trivially true.

Let $n > 1$, and assume that the inequality holds for $1, 2, \cdots, n-1$. If n is even, we write $n = 2m$, and by the induction assumption $s_m \leq (k-2)2^{k-1}+1$, since $m \leq 2^{k-1}$. So, applying the first relation of (52) we have

$$s_n = 2s_m + n - 1 \leq (k-2)2^k + 2 + 2^k - 1 = (k-1)2^k + 1.$$

If n is odd, we write $n = 2m + 1$, and again by the induction assumption $s_m < s_{m+1} \leq (k-2)2^{k-1} + 1$, since $m + 1 \leq 2^{k-1}$. Therefore

$$s_n = s_{m+1} + s_m + 2m \leq (k-2)2^k + 2 + 2m$$

$$= (k-2)2^k + n - 1 \le (k-1)2^k + 1.$$

Hence, relation (54) holds for all n satisfying $n \le 2^k$.

Taking logarithms of both sides of (54) gives us the asymptotic behaviour $s_n = O(n^2 log\ n) = O(n\ \ell n\ n)$.

References

[1] Graham, R.L., Knuth, D.E. and Patashnik, O. Concrete Mathematics, Addison-Wesley Publ. Comp., Reading, Mass., 1989.

[2] Grimaldi, R.P., Discrete and Combinatorial Mathematics, Addison-Wesley Publ. Comp., Reading, Mass., 1989.

[3] Liu, C.L., Introduction to Combinatorial Mathematics, Mc Graw-Hill Book Comp., New York, 1968.

[4] Maurer, S.B. and Ralston, A., Discrete Algorithmics, Addison-Wesley Publ. Comp., Reading, Mass., 1991.

[5] Papadimitriou, C.H. and Steiglitz, K., Combinatorial Optimization: Algorithms and Complexity, Prentice Hall, Inc., Englewood Cliffs, N.J., 1982.

[6] Reingold, E.M., Nievergelt, J. and Narsingh Deo, Combinatorial Algorithms, Prentice Hall, Inc., Englewood Cliffs, N.J., 1977.

[7] Truss, J.K., Discrete Mathematics for Computer Scientists, Addison-Wesley Publ. Comp., Reading, Mass, 1991.

Refs. [2], [3], [4], and [7] are of an introductory nature, whereas the other books are of a more advanced level.